BRONZE CASTING
A Manual of Techniques

Bronze Casting

A Manual of Techniques

Guy Thomas

The Crowood Press

First published in 1995 by
The Crowood Press Ltd
Ramsbury, Marlborough
Wiltshire SN8 2HR

British Library Cataloguing in Publication Data
A catalogue record for this book is available from the British Library.

ISBN 1 85223 938 7

Picture Credits
All photographs by Guy Thomas and Barbara Cheney
All line-drawings by Annette Findlay

Acknowledgements
This book would not have been written without dedicated help in all areas of
the production of this work, making sculptures, photography and text
preparation. My eternal thanks to Barbara.
  This book is dedicated to the bronze workers of the ancient world whose skill
in casting using relatively primitive equipment is inspirational to the modern art.

Typeset by Dorwyn Ltd, Rowlands Castle
Printed and bound by The Bath Press

Contents

Introduction

How many of us walk round a gallery or museum looking at sculpture, both ancient and modern, reading the labels – 'Iranian finial 850–650 BC, bronze' or 'Barry Flanagan, *Hare*, 1994, bronze' – but without even wondering what bronze is, and how a sculpture ever came to be cast into bronze. I think that most of us accept without further consideration the fact that a sculpture is made of bronze; though we can perhaps be forgiven for thinking that bronze is a material that sculptures just happen to be made out of, in the same way that sculptures are made out of stone or clay.

The truth is that bronze has to be cast in a molten state using a series of techniques that date back to approximately 2000 BC. Archaeologists have estimated that by 7000 BC copper was being hammered and rolled, and that by 5000 BC it was being melted and poured into open moulds of stone and earthenware. This led to the discovery that if copper was mixed with arsenic the metal became stronger and more malleable. In turn arsenic was replaced by tin, and bronze came into being; bronze consists of about 80 per cent copper, 1–2 per cent tin and zinc, and the rest is made up of many elements in small amounts, including lead. The other main copper alloy is brass, which is a mixture of copper and zinc. Apart from sculpture, bronze was used mainly for weapons and armour as well as domestic items such as pots.

LOST WAX TECHNIQUE

The main technique for casting bronze today is the 'lost wax' technique, known as *cire perdu*. The Egyptians were using *cire perdu* as early as 1570 BC. Put simply, a sculpture made in wax was encased in clay; when baked, the wax melted away and the clay became the mould for the bronze to be poured into. The clay mould would then be broken open to reveal the solid bronze cast.

Today the investment method carries on the same technique as that of ancient times. The process involves taking a sculpture which has either been modelled in or cast into wax, and attaching a system of solid wax rods called runners and risers. A solid wax sculpture must not exceed 1in (2.5cm) in thickness; otherwise it will have to be cast hollow utilizing a core (an inner mould) which is made out of the same plaster mixture and held in place with pins. The runners and risers will become hollow channels when the wax melts away in the kiln, allowing the bronze to flow in and the gases to rise up when the bronze is poured.

The wax is then covered with plaster, finally being built up into a plaster mould with the wax inside. This mould is then put upside down into a kiln and baked for approximately three days: the wax melts out within the first few hours, and carbon deposits are burnt out at a later stage. The mould now has an empty cavity (which was the wax sculpture) with its runners and risers; it is then buried in a sand pit, the sand being packed around it to support the pressure of the bronze when it is poured. The bronze is smelted down, and in its molten state is poured into the mould filling the empty cavity so that what was the original wax sculpture now becomes a bronze one.

After the bronze has been poured and has cooled, the plaster mould is broken away and the bronze system of runners and risers is cut off. The bronze then has to be worked on to remove any casting defects, and to obtain the original surface that the wax sculpture had. Tools are used to reproduce the original surface where that surface was lost to the runners and risers.

Bronze is a very sensitive material that will pick up every single detail, including fingerprints that were left on the wax; for this reason alone it is an extremely exciting material to have a sculpture cast into. As well as wax, other materials that are light enough to burn out in the kiln with the wax can also be used; these may include some types of wood (for example balsa wood), paper, cardboard, polystyrene, fabrics and found objects such as dried flowers and fruits. This further adds to the exciting range of possibilities for bronze sculpture.

Bronze sculpture by Guy Thomas. This sculpture was modelled out of wax and made directly without an armature. The torso was modelled using large lumps of wax, sticking these together in such a way as to create the rugged expression of the character. The hat was made out of a piece of towelling which was thickened slightly with wax: in its final bronze form this has been replicated exactly. This sculpture was cast using the 'lost wax' method; 9 × 9 × 4in (23 × 23 × 10cm).

THE THREE METHODS OF CASTING

There are three main methods of casting in bronze: the investment method, the ceramic shell method, and sand casting. The investment and ceramic shell methods both use the lost wax process. The investment method is that of utilizing a plaster mould, as just described; the ceramic shell method consists of a process of dipping the wax sculpture into a ceramic mixture to build up a thin layer or shell. This is then baked in the kiln, where the ceramic coating becomes very hard. The wax is again burnt out and lost in the same way as with the investment method, and the empty cavity filled with molten bronze.

Sand casting is the process of placing an object, normally wood or metal, into a sand mixture in a box that is in two halves. The object is buried half-way in one half, and the sand in that half is frozen with CO_2 gas. The second part of the box is attached to the first half and sand is packed into the second half and frozen. The box can then be split in two and the object removed, leaving an empty cavity in the sand which can then be filled with molten bronze. The bronze is then poured in via a runner and vent system. This is the quickest method for casting bronze, but it only suits geometric forms.

MAKING SCULPTURES

Casting sculptures into bronze is very traditional. This book is intended to demonstrate as many techniques as possible for making a sculpture that will be cast into bronze: there are many quick and easy methods, especially when using wax directly, as well as longer and more complicated methods of plaster casting. Chapter 16 deals specifically with what actually happens to your sculpture at the foundry when it is cast using the lost wax (investment) method.

I hope this book will inform and unravel some of the mystery that surrounds the process of bronze casting; it is one which is quite difficult to grasp unless you have gone through the casting side of it yourself. Unfortunately most of you will not have this opportunity, but I hope you will have a better understanding of it from reading this book. The art of bronze sculpture is timeless; perhaps you will be able to add something of value to this most ancient and modern art form.

PART I
Methods of Working

The first chapter tells you what sort of facilities you will require to get started in making sculptures for bronze casting, and what tools are essential.

The next four chapters introduce the main materials for making sculptures to be cast into bronze. These materials all have different qualities, and which one you choose to work with will determine the form your sculpture will take. It makes no difference whether you are modelling, carving or making a construction using clay, plaster or wax: whichever material you choose to use, you should fully understand what it can and can't do. Some are more direct to work with than others. For instance if you want to model a figure to be cast into bronze, you may choose to work with clay; this will require an armature, and once finished it will then need to have a plaster waste mould made in order to make a plaster positive, from which a rubber mould can be made to obtain a wax ready for the bronze casting. However, if you were to use wax to model your figure, none of this would be necessary because the wax-modelled figure can go to the foundry as it is.

This should not put you off working in clay: clay is the traditional material for modelling portraits and the figure, it is one that people may know well, and it is well liked as a sculptural medium. But just because you may be familiar with one material should not in any way stop you from experimenting with another. With wax you can do your modelling in exactly the same way as you would with clay, but it also opens the door to a whole variety of new techniques to which perhaps you haven't given serious consideration before. Indeed, wax is the traditional material for making sculptures to be cast into bronze.

There are lots of techniques to explore, and the chapters which follow will, I hope, help you to do just that: after reading them I hope you will have a better understanding of the materials, and more confidence in their use. By pushing them around you will find out for *yourself* the possibilities and limitations of each material. Moreover, you should exploit each one to the full, and by doing this you will be working towards making a better sculpture.

Somewhere to Work

STUDIO SPACE

Finding somewhere suitable to undertake the practice of sculpture can be a problem. Ideally a studio space should have plenty of natural light, be dry, have plenty of room and – perhaps most importantly – somewhere where you can make a mess within reason. It may be more convenient if it is on ground level, too, so you avoid having to heave heavy bags of clay and plaster up flights of stairs.

Having a real studio allows you to concentrate your focus of attention solely on your artistic practice, surrounded by tools and materials, in an atmosphere of artistic industry. This can add to your sense of purpose, changing what may start off as a hobby into more a way of life, changing your perception of how you view objects and people in the real world. If you cannot afford to rent an artist's studio, then several options are open to you. A garage can make a very good studio space, especially in the summer; although in winter this could be a different story, with bad light and cold to contend with. However, with artificial light and a couple of fan heaters installed, a garage could become quite cosy; it also has the advantage of being on ground level with easy access to remove waste materials, and you can quite permissibly make some noise when necessary. A shed or outbuilding could also be adapted to a studio space; or a spare room could be used.

If none of these options is open to you, there is no reason why a kitchen cannot be used. Obviously, there are disadvantages to using a kitchen: possible interruptions, mess, and having to clear away tools and materials at the end of each sculpting session. However, if you can put up with these minor inconveniences, and use plenty of protective coverings such as dust covers and plastic sheeting around the working area, then all the procedures described in this book could be undertaken in a medium-sized kitchen. If you are naturally a clean, neat worker and tidy up as you go along, then working in a kitchen will pose no real problem. In fact, most of the procedures described can be accomplished working on a table in the comfort of your own home.

TOOLS AND MATERIALS

Most of the equipment used in a sculptor's studio can be bought cheaply or may be donated to you. Some form of work surface is essential, perhaps an old table; some old shelves to make into modelling boards; and some scrap timber retrieved from a skip to use either for armature materials or as props. It is very useful to have a central table in your studio; this could be a modelling stand which has the advantage of height adjustment and a turntable. A banker, normally used for carving, is also good because it is higher than a normal table and is extremely sturdy; but they tend to be expensive to buy. Some form of turntable is a must for all modellers: this can be either a wooden one, or an aluminium whirler as used in ceramics. Whirlers are cheaper and light in weight.

A number of cheap buckets and bowls will come in useful, especially with clay modelling and plaster casting procedures. Most of the tools needed can be bought at a car boot sale, for example hammers, mallets, old chisels, files, a power drill and screwdrivers; an old saucepan and old dinner knives will also be of great value when working with wax. More detailed lists of specialist tools are given at the beginning of the appropriate chapters.

The basic materials needed can be acquired either free or very cheaply: wood, paper, card, polystyrene and clay are all materials which are quite inexpensive or can be acquired in scrap form. Wax, plaster and the various rubbers used in mould making are more costly items, but can be bought within the restraints of a tight budget. You will also find that these materials will last a long time if you are working on a smaller scale of sculpture. Wax and clay can be re-used over and over again, if you look after them properly.

A table or work surface in a kitchen can easily be modified to facilitate a studio space. With plenty of protective covering such as polythene sheeting any amount of mess can be contained.

It is better to buy clay, wax and plaster from the specialist companies who supply them as you will have a greater variety of types to choose from, and the price will be cheaper as you will be able to buy in larger quantity. For any of the procedures incorporating the use of plaster, you will need to buy a minimum of approximately 25kg (55lb).

It is very useful to have a good camera to record the stages in making a sculpture. A compact camera with a flash is fine, but I would recommend a manually operated SLR camera so that you can control the depth of field and compensate for lighting conditions. Always take your light reading off the sculpture itself, not its surroundings.

Armature Making

For the purposes of clay modelling, a basic understanding of armature making is essential. A model needs an armature to support it in exactly the same way as a building needs a foundation, beams and girders if it is to stay up.

THE STRESSES OF CLAY

Unlike wax, clay has no tensile strength and is a very heavy and dense material which needs internal support. When sculpting any form of figure or animal out of clay you cannot escape the laws of gravity unless using a welded steel armature. It is therefore important to remember the principle of a tripod support on three legs when building an armature, especially when undertaking forms that stick out in several directions, for example the arms or legs of a figure. As humans we are able to stand because our muscles hold us in balance; a sculpture, however, needs the firm, fixed support of an armature.

As clay is added to an armature, more stress is put on that armature: it must therefore be absolutely stable, as any looseness in its construction will become more evident as the sculpture is built up. If the armature is not sound and the weight stress increases beyond its holding power, it could be disastrous. It is therefore imperative that all props on a modelling board are securely screwed, and that top props are tightly bound. Moreover, a top-heavy form, especially one that extends sideways, will cause an incredible amount of sideways stress to a central prop, and it may be necessary temporarily to add another prop under such a form to give extra support.

MODELLING STANDS

Armatures can be made out of wood, steel and wire, and normally consist of a modelling board with some form of central prop; a form can either be built around this, or an armature can be built separately, but supported by this prop (see diagram).

Modelling stands can be bought. A stand consists of what is known as a back iron – an upright piece of steel bar bent at a right angle at the top – secured to a base board. The back iron can be used to support a sculpture at the back, or it can be extended: there are screw holes at the top of the back iron to which a separate armature can be attached, and you can add wood or modelling wire, for example, to profile a pose of outstretched limbs. A sliding armature support can be bought, too, and by varying the sliding arm either up or down or on a slant it is much easier to fashion a figure or animal in any pose. It is therefore ideal for modelling figures in action or birds and other animals.

Making your own Modelling Stand

You can make your own modelling stand by securing a central prop or back prop to a board by screwing four metal shelving brackets to a piece of timber. Other pieces of wood or wire can then be attached to this prop to form a secure armature. It is important that armatures retain some flexibility; you need to be able to change them quite easily by bending a limb here or an arm there. Wishing to change the pose in some way is a very common necessity in the modelling process. Aluminium modelling wire is ideal for this, and can be bought in a variety of thickness.

If you are building an armature for a simple figure with no outstretched limbs – for instance, someone standing with his feet together and his hands behind his back – then you could use wood bulked out with polystyrene built directly over a central prop. Polystyrene is a very useful material, and is easily obtainable as the packaging material of consumer goods. Also pieces of old dexion (shelf-building material) can be used in conjunction with a home-made wooden armature, as they can be easily screwed to a wooden prop, and other pieces can be bolted to each other to profile a pose.

For smaller sculptures an armature can be made out

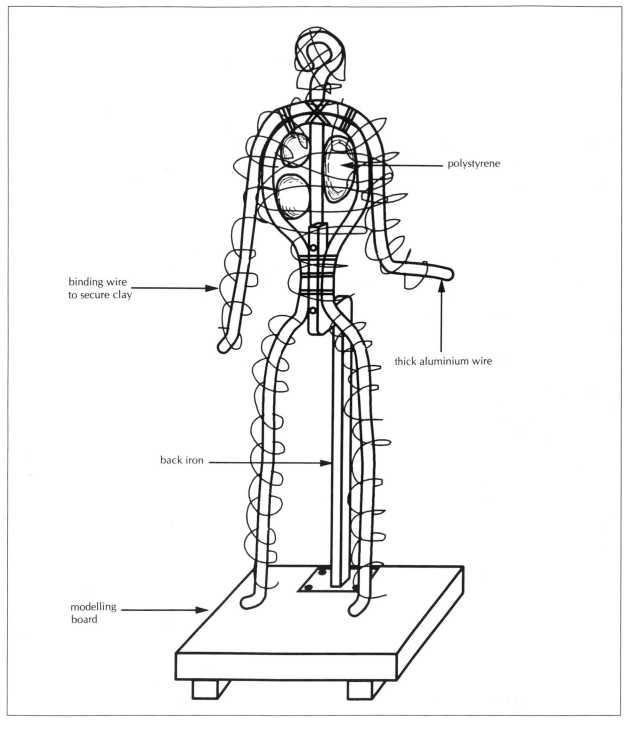

polystyrene

binding wire
to secure clay

thick aluminium wire

back iron

modelling
board

Modelling board with back iron supporting an aluminium wire armature of a figure. Note the binding wire wrapped around the thicker armature wire to key the clay. The armature can be bulked out with pieces of polystyrene.

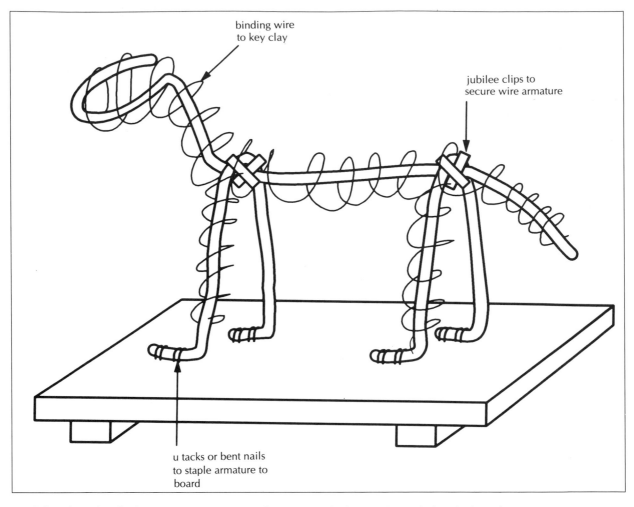

binding wire
to key clay

jubilee clips to
secure wire armature

u tacks or bent nails
to staple armature to
board

Modelling board with aluminium wire armature for an animal. The wire is stapled to the board with U tacks or bent nails. The legs are secured to the body with jubilee clips. Note the binding wire wrapped around the armature to key the clay.

of aluminium wire, which has been bent to a right angle and U-tacked onto a modelling board. Other pieces of wire can be then lashed to this central piece (see diagram above).

Armatures made of welded steel are by far the best for life-size sculptures as they really can defy gravity, as long as the steel base is large enough. The only problem is that they are heavy to move and cannot be changed so easily; it is important therefore to have decided on the exact positioning of the pose beforehand.

When building large armatures completely out of wood, always use screws to join pieces together, then turn the joined wood over and screw them again from

the other side to ensure a strong fixing. For joining pieces of aluminium wire together, small jubilee clips can be useful to ensure a good join.

Getting Ready to Put the Clay on

Whatever armature material you use it is important to have a surface on which clay will stick firmly. To achieve this on smaller sculptures, wrap binding wire around the areas where clay is to be added; this will 'key' (stick) the clay firmly to the armature and prevent it slipping off.

When you have built your basic armature structure,

it needs to be bulked out to form the general shape of your figure. It should be only slightly smaller than the intended size of the sculpture, however; you want it to be near the surface of the sculpture to avoid unnecessarily thick areas of clay, and since it will determine the shape and size of the sculpture, it is worth spending the time to get it right. If you add a large amount of clay to an outstretched limb or a top-heavy form then you may need to bring in an additional prop, placing it underneath for support; it can be removed whenever you like.

For larger sculptures, chicken wire is very useful in bulking out the shape: it can be wrapped around, bent and manipulated to get the basic form for you to start modelling on. You can then bulk out the chicken wire with polystyrene or tightly screwed up pieces of newspaper. Clay keys very well to chicken wire. Old pallets are very good for constructing larger armatures.

Working in Clay

Clay as a material has been used by people since pre-historic times for both functional and sculptural objects. Once the clay has been manipulated to make an object it is allowed to dry out thoroughly and then fired to transform it into a harder and more permanent material. Functional objects are usually glazed to colour the clay and seal its porous surface.

Clay is likely to be one of the first materials most sculptors ever use. At some time or other most people have used it, perhaps at school to make simple ceramic objects, so it is probable that nearly everyone will have some familiarity with the material.

CLAY SCULPTURE

For sculptural purposes clay is used as the main modelling and casting material; it can also be used for making 'maquettes' (small studies for a larger sculpture). This chapter deals with techniques you can use to make sculptures, but clay is also a very important and flexible aid in the casting process where it does not become the end product. For any casting process, therefore, it is an invaluable material with many uses: for instance, it is used to build clay walls in making a plaster waste mould (see Chapter 11); to contain plaster when making a plaster cast (see Chapter 17, Making a Clay Relief); it can be used to make press moulds (see Chapter 4); and it is essential for making a rubber mould, where it is used to make a 'clay blanket' (see Chapter 13).

Clay has both advantages and disadvantages. It is easy to manipulate, cheap to buy, it can be worked and reworked easily, and provided it is properly looked after, can be re-used over and over again. On the down side, it is essentially an inert material, and if you want to make a largish form it needs the support of an armature. It has to be kept wet and worked wet; it must not be allowed to dry out while you are still working on a clay sculpture as you cannot add wet clay to clay which has dried out. When it dries out it is prone to crack and flake, and it needs careful handling as it tends to break very easily.

Casting Clay Sculptures

Casting a sculpture which has been made in clay is a long and rather laborious process. It is normally cast by making a plaster waste mould – a plaster shell – which is often in many parts. Once the plaster mould has been taken away from the clay sculpture it is reassembled independently and will contain a cavity which was the original sculpture; now a material such as plaster or cement can be poured in. When the material has set, or 'gone off', the shell or case is then chipped off to reveal the sculpture reproduced in the new material. Chapter 11 describes in more detail how to make a plaster waste mould for a clay head.

A more direct material to work with is wax. However, at first most people find clay easier to work with than wax.

Types of Clay

There are many types of clay, all of which are variations of the three main types, which are fired at different temperatures:

- Earthenware clay is a lower-fired clay which remains porous unless glazed.
- Stoneware is a hard, non-porous clay which is fired at a higher temperature. It is intermediary in character between earthenware and porcelain.
- Porcelain is a ceramic material in which felspathic quartz and sericite rocks and pure kaolin are fired to give a hard, translucent white body.

The most common clay used in sculpture tends to be a grey clay known as School Buff. It is smooth and contains about 10 per cent grog (ceramic material), which gives it more strength when modelling with it. You can also use another grey clay called Industrial Crank which has a much higher percentage of grog and is good for

TOOLS AND EQUIPMENT

A set of clay modelling tools You can buy either steel, plastic or wooden modelling tools. I would suggest using steel ones (buy them individually as they are expensive); they are extremely useful and can also be used when working in wax. You should also buy a set of either plastic or wooden modelling tools as these undoubtedly will be used at some point during the modelling process.

Old kitchen utensils As an alternative to buying steel modelling tools, you can use old knives and spoons to obtain good results.

A wooden rolling pin This is essential for rolling out slabs of clay; or you can use an empty wine bottle as a cheap alternative. Two slats of wood approx 18in (46cm) long and ¼in (6mm) thick are useful for rolling out clay to an even thickness.

A water sprayer A cheap plastic sprayer is very useful to keep clay wet, and for use in the casting process.

Hessian or polythene sheeting Roll clay on this. Polythene should be oiled first, either with washing-up liquid or cooking oil. Hessian should be soaked in water and the excess wrung out before using it to roll clay onto. It does, of course, leave an impression of its surface on the clay, so may not be suitable if you are constructing a sculpture. Plastic shopping bags will come in very handy to cover your sculpture.

Turntable This is very useful for working on three-dimensional sculptures.

A wire This is to cut clay cleanly. You can buy a ready-made wire, but to make your own simply take a

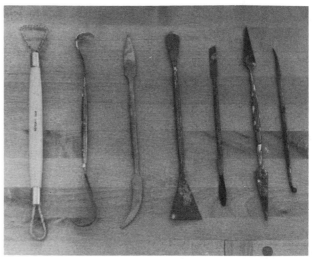

A set of steel modelling tools. It is not necessary to buy a whole set; it is better to buy them individually as you will find yourself using only one or two.

length of fishing wire about 18in (46cm) long and wrap each end around a wooden peg made out of dowelling.

---

building large sculptural forms; it is not so good for fine modelling, however, as it is very sandy to touch.

There is also a red earthenware clay available which has exactly the same properties as School Buff, is good for modelling, roughly the same price and is perhaps a more appealing colour to work with.

SLABBING CLAY

Slabbing clay is a technique used mainly in ceramics, where objects are hand-built out of thick sheets of clay. In sculpture, this technique can be used in constructing more abstract sculptural forms and for cutting out shapes to use with modelled forms or as details. When clay is slabbed it is probably in its strongest structural form and can therefore often be worked without the use of an armature. Clay worked like this is also in its

neatest form, so this technique may suit some artists who like to work in a more minimal constructivist way.

In this book, slabs are mainly used to build what are known as 'clay walls': literally, walls of clay used in the casting process to divide areas or to contain plaster when it is poured.

A slab of clay is a good base for making a relief (see Chapter 17).

Wedging

If you want to have a slabbed sculpture fired, it is imperative to 'wedge' the clay first. 'Wedging' clay is the process of removing air from the clay to prevent the danger of an object exploding in the kiln. There are several ways to wedge clay.

The way I do it is to take a lump the equivalent of two handfuls, and cut it in half with a wire or knife;

slap one half onto the other, and then roll the whole lump slightly to form a curved underside. Slapping the clay down forces the air out, and you can be quite heavy-handed when wedging. Pick up the lump from underneath in one hand and slap it down onto the work surface, turning it a quarter turn as you do so. Repeat this process for at least ten minutes. Shape your clay into a block and cut slices into it using a wire to check that there are no air bubbles remaining.

Do your wedging on a porous surface – a plaster block is ideal – because if the clay is very wet this will help the excess moisture to be extracted. If you are constructing with slabs of clay they must not be too wet as they will bend too easily and your sculpture will be in danger of collapse. If the clay is still too wet after you have rolled the slabs out, lay them on a board one on top of the other with a layer of newspaper in between, cover loosely with polythene, and leave them to dry out *slowly*.

Rolling Out

To roll out a slab, simply take a lump of clay and place it onto the oiled plastic sheeting or length of damp hessian. To ensure an even thickness, place two lengths of wood, both of the thickness you require, on either side of the clay when rolling out. The thickness will depend upon the job in hand but if you intend to have your sculpture fired, it should not exceed ½in (13mm). Once rolled out, the slab can be cut to shape.

Joining Slabs

To join slabs together, the surfaces to be joined should be 'keyed' with a knife point or another pointed implement. Use your fingers to apply some clay slip (a solution of clay and water) to both surfaces. The slip is the 'glue' which fixes them together; the slabs should be supported while they are adhering.

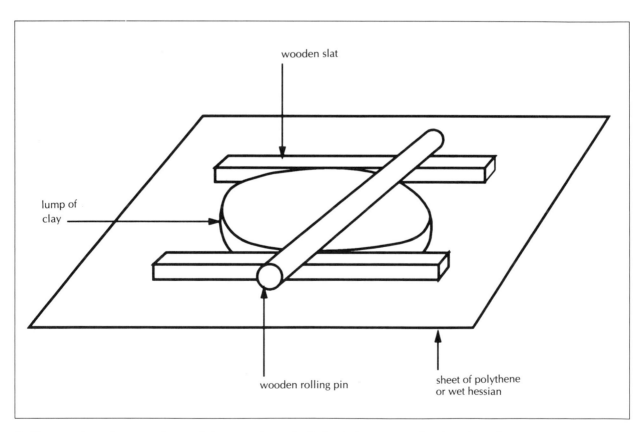

Rolling out clay: place your lump of clay on a sheet of oiled polythene or wet hessian; then place two wooden slats, as thick as you require the slab of clay to be, on either side of it. The clay can now be rolled out evenly.

MODELLING CLAY

Modelling clay is essentially a straightforward process of adding and subtracting, putting it on and then quite often taking pieces off again until a satisfactory form is achieved. With clay, unless the sculpture is very small, it needs to be built over an armature, as clay is a very heavy and dense material. It has no tensile strength of its own, and because of its properties it creates a lot of weight stress, so a sound armature is essential to avoid the disaster of collapse (see Chapter 2).

A small sculpture, however, can be made with just a simple armature fashioned from a piece of aluminium wire bent to a right angle and U-tacked onto a modelling board; other pieces of wire can be lashed to this central piece (see diagram, Chapter 2).

CLAY PRESS MOULDS

Clay can be used as a mould material in its own right. Simply by making indentations, or by carving and modelling negative forms in the clay, you can fill theempty cavities with other materials such as plaster, wax and even cement. Chapter 19 deals with such a project in detail.

Clay press moulds can be used in quite a different way as well. Instead of taking an object to the clay and pushing it in to make an impression, a slab of clay can be taken to an immovable object such as a building, tree, statue, stone monument, rock and so on, and used to take a quick, accurate impression. I have heard stories about certain young artists actually using a small bed of clay with clingfilm over the top of it to take impressions of parts of famous works of art in museums and galleries. Obviously I would not advocate such a practice here. But the same methods can be applied to more accessible – and legal – subject matters, and you can easily use clay to take an impression from a sculpture you have already made out of a hard material.

To take such an impression, use clay that is not too wet. To test this, push your finger into the clay: if the clay sticks to your finger it is too wet – your finger should be able to pull away freely without any mess. Use a large handful of clay, and either pat it flat on one side, or roll it out to a slab about 1½in (38mm) thick. Smack the clay firmly onto the subject matter: use your knuckles to really push the clay against the object. Then carefully prise it away. If you are taking an impression from something in the landscape or on site, lay the clay impression on its back in a container to be transported back to your studio. A positive can then be cast either in plaster or wax.

By taking such impressions you can achieve some very interesting textures or details. You could incorporate these into a wax or plaster sculpture, or even cast them into bronze as a relief.

CASTING USING WAX

To cast a sculpture by making a wax mould is a very direct and quick process. Chapters 10 to 12 deal with modelling a portrait head and casting it into plaster by using a plaster waste mould, which is a more complicated process. However, the wax process demonstrated on p. 26 is more straightforward, although it is only suitable for smaller sculptures with a simpler form.

DEMONSTRATION: MODELLING A SMALL CLAY FIGURE

This figure is only 6in (15cm) high and is based on a sculpture by Matisse. It is worthwhile trying a small sculpture first before tackling a large one, to familiarize yourself with the techniques involved in modelling and casting into plaster. To make this sculpture I made a simple wire armature and started modelling at the bottom, building up the structure over the wire. I cut the base from a slab of clay, and fashioned the legs by pushing small lumps of clay onto one another, building up the rough form to about 3in (7.5cm). Once this was established, a piece of wire was inserted and the torso was continued up over the wire. This was built up by adding more small lumps, pushing them into one another to make a continuous surface. I pushed the clay around until I got the form I wanted, then smoothed it over and, using a modelling tool, cut away again until I was happy. The legs were left in a rough form whilst the rest of the body was built up over the wire. The arm and head were made in a similar way, building up with small lumps of clay.

When you work with clay in this way you will notice the different effects you can obtain: building up with small lumps creates an expressive quality; smoothing over gives a more finished look; and carving away gives a raw look to clay, and in any given sculpture you are likely to use all these techniques. You can use your modelling tools to mark the clay by scribing in lines or detail. There are no real rules to modelling clay – it's a question of finding your own way of working with it.

On such a small sculpture as this you can quickly establish the forms, thinking about what is happening around the other side. If you work on a small turntable you will be able to keep turning the sculpture around with ease, examining the developing structure in the round all the time.

A small sculpture like this can easily be cast into plaster using a simple process via wax.

Building up the clay over a piece of wire: add lumps of clay and build the form from the bottom upwards.

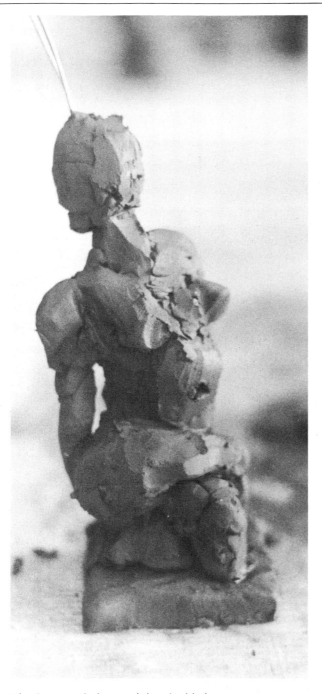

The figure with the rough head added, establishing the form of the figure.

Starting to refine the form: using the modelling tool to cut away slices of clay.

DEMONSTRATION: CASTING A SMALL CLAY FIGURE

In the example shown here the arm would be more easily cast if it were closed into the body, with no gap between arm and body. However, I chose this pose for the figure deliberately, to demonstrate what can be achieved by a simple process: as you will see, the wire armature proves indispensable, acting as a handle to the sculpture.

First you will need to melt some wax in a saucepan, as described in Chapter 5. When the wax is melted but is not too hot you can either dip the sculpture into it, or brush the wax onto the model. Brush it evenly all over, letting the wax fall gently off the brush rather than brushing from side to side continuously.

After each dip or layer of brushed-on wax, cool off the sculpture in cold water to harden the wax and allow the next layer to adhere. Continue this process until the wax coating is approximately ¼–⅜in (6–9mm) thick, then place the whole piece in cold water for about five minutes to ensure the wax has hardened. The mould can then be cut very carefully with a craft knife straight through the middle, all the way down until it is completely in two halves.

Remove the clay and wash out the moulds with cold water. Trim off any excess wax around the bottom, locate the two halves back together and seal them by using the hot knife technique (see Chapter 5) all the way around the join.

The mould is now ready to fill with plaster (see Chapter 4, Mixing a Bowl of Plaster). It can be propped upside down: in this case I simply put it in a tea mug to support it and filled it level with the bottom. After pouring in the plaster, tap the mould lightly with a wooden spoon to ensure that any air bubbles rise to the surface.

Leave the plaster to set for at least two hours. The wax can be removed either by gently burning off with a flame torch, or by placing the mould in hot water and gently peeling the wax off in sections. Carefully cut three-quarters of the way through the wax to achieve this.

A line of flashing may be present around the plaster sculpture where the two halves of the wax mould were joined, and this can be cut and scraped away. Any other marks left by the casting process can be filed or filled with a small mixture of plaster.

The completed wax mould. Note the wire sticking out of the top which you can use as a handle.

The empty wax mould with the two halves sealed back together.

The completed plaster figure; 6 × 2½in (15 × 6.5cm).

Working in Plaster

Plaster is used in sculpture mainly as a casting material, allowing an inert material such as clay to be turned into a more permanent one. However, plaster can also be used for constructing with and modelling directly; it can also be carved, though perhaps less successfully.

The main constituent of plaster is gypsum, a mineral consisting of hydrous sulphate of lime; it is often called plaster of Paris due to the large deposits of gypsum that were found in the hill of Montmartre. It has been used as a casting material since the time of the Egyptians. Gypsum needs to be heated to about 100°C (212°F) to form a powder, which when recombined with water will set hard.

When you mix and use plaster it starts off being runny (the consistency of single cream), and after about ten minutes thickens to a cheesy state (the consistency of cream cheese). Soon after this the plaster sets, or goes 'off', as it is called, and if you place your hand on its surface at this stage you will feel warmth. This heat is equivalent to that which was needed to convert the gypsum to plaster. Plaster is 'cured' when it dries out and reaches its hard state.

TYPES OF PLASTER

There are many types of casting plaster to choose from, each type relating to its hardness. Ordinary Fine casting plaster is relatively cheap to buy and good for most general sculpture jobs, especially for construction and the rougher aspects of mould making. However, the better plaster to buy is called Superfine casting plaster, which may be sold under a different name, depending on which supplier you obtain it from. Superfine is of a much better quality than Fine casting plaster; once it has cured it is harder, and it picks up intricate details better, but it is more expensive. Herculite is a very hard plaster which is used with grog (ceramic material) for the first coat of an investment mould in bronze casting.

To obtain any form of casting plaster you will have to buy it from a specialist dealer; it is unlikely that you will obtain it from your local DIY store. It should be noted that there is a vast difference between walling plaster as sold in DIY shops and casting plaster, so do not attempt to use walling plaster unless you are constructing with it.

Plaster Care

Providing you keep your casting plaster wrapped in plastic and dry it will last a long time, although plaster does have a shelf life of about a year. Working with plaster is a messy business even if you are a neat worker; it tends to get everywhere, so always wear old clothes and protect your working area with plenty of newspaper or polythene sheeting.

To wash plaster off your hands, use a bucket of warm water, *not your kitchen sink* as plaster very easily blocks drains. In the bucket of water it will sink to the bottom and can be disposed of more easily. To clean a bowl which has been used to mix plaster, wait for the plaster to go off and then simply flex the sides of the bowl to release the pieces of dried plaster. If the sides of your bowl do not bend, bang the bottom on the floor. It is better to wait for the plaster to dry and do this rather than washing out wet plaster.

You may wish to wear a face mask to cut down the amount of plaster dust you breathe in.

MIXING A BOWL OF PLASTER

Try to acquire a proper plaster mixing bowl which is

Mixing a bowl of plaster. Sprinkling the plaster round the bowl.

more flexible than, say, a washing-up bowl. Plaster suppliers sell these bowls and it is worthwhile buying several. A normal washing-up bowl will suffice, but will be harder to pour from and to clean.

Fill your bowl slightly less than half full with cold water. Take a handful of plaster and sprinkle it over the surface of the water as evenly as possible; if lumps fall into the bowl, remove them. Continue to sprinkle handfuls of plaster into the bowl until you see an island of plaster form in the middle; there is no need to rush, as there is plenty of time before the plaster will go off. Then add one or two more handfuls of plaster.

Mix the plaster by placing your hand in the bowl and moving it from side to side, working from the bottom up. You should not mix plaster with both hands as though you were mixing pastry, nor should you mix it with a wooden spoon. Make sure there are no lumps in the mixture.

Once the plaster is mixed, check to see if it is of the right consistency by scooping some up in your hand then letting it fall back into the bowl: the consistency should be of single cream – not totally liquid, but not like yoghurt. Plaster goes through various stages in setting. Its initial runny state is the right consistency for flicking when you are making a plaster waste mould (see Chapter 11). You will notice it getting thicker; after about ten minutes it will get to a 'cheesy' state when it has the consistency of thick porridge. This is the right stage for modelling or constructing with it, though you have to work very quickly with it in this state as it dries soon after (see the demonstration at the end of this chapter, Making a Plaster Sheep).

Once dry, the plaster will go through a 'warm' stage which you will feel very distinctly and is quite pleasant, especially in winter after having your hands repeatedly in cold water. Its final 'curing' time – its proper hard state – is much later, although once set initially you can work on it either by using a surform, or

DEMONSTRATION: MODELLING A PLASTER SHEEP

This example is very quick and direct, and it exploits the material in a very successful way. I used a simple armature of polystyrene shaped roughly into the body of a sheep. Into this I stuck four twigs for the legs. I applied the plaster in its cheesy state, fixing the legs in position first and then modelling the top and the head of the sheep. You have to work very quickly; there are probably only about five minutes in which to work before the plaster becomes too dry. I managed to complete this sheep in one go, but you can add more plaster if you need several attempts, providing the plaster is wetted down so the new layers can adhere properly. I did not smooth or refine the surface of the plaster, as the texture of the plaster when applied loosely emulated that of the sheep's wool.

If you wish to add a new layer of plaster, the surface of the layer already applied should be 'scored', that is, marked with scratched lines to ensure that the subsequent layer adheres properly. 'Wetting down' means actually soaking the plaster form in water, although this is only necessary if the previous plaster layer has completely dried out.

This simple little sculpture would look terrific cast into bronze, so try some other examples and experiment with your forms!

The polystyrene armature with four twig legs for the plaster sheep.

The completed plaster sheep (side view);
6½in × 9 × 4in (16.5 × 23 × 10cm).

The completed plaster sheep (back view).

by carving, or removing it from its mould. A surform is a two-handed tool similar to a long cheese grater, used to smooth plaster and refine a form. You can also buy plaster rifflers, one-handed tools used to smooth plaster.

CASTING USING PLASTER MOULDS

Plaster casting is often a complex business as every sculpture made to be cast has its own individual problems, and therefore each one may have to be tackled differently. A plaster mould of a sculpture is made in order to facilitate a cast. Briefly, the sculpture is divided into sections and a mould is made of each in turn, so that when all the pieces are joined together they make a complete mould; the original sculpture can then be reproduced in a new material, for instance into plaster or cement.

Chapter 11 describes in detail how to make a plaster mould from a clay head. This is essentially a straightforward process as the mould can be tackled in just two halves. On a more complicated sculpture such as a figure or an animal, the mould may have to be cast in many sections or 'caps' so the clay can be removed from the plaster casings. On a larger, more complicated sculpture – for instance a figure which has protruding limbs – the process of capping can become extremely complicated, and the caps themselves would need reinforcing on the outside with lengths of iron or wood.

On a large sculpture the cast would need to be hollow and would therefore require the added complication of a steel armature to support it. It is beyond the scope of this book to discuss this; here, we are dealing with more elementary processes. There have been several books written specifically on the casting process if you need more information.

MODELLING AND CONSTRUCTING WITH PLASTER

Plaster can be used for constructing very large hollow sculptures which may later be cast into bronze. Many of the large bronzes by Henry Moore were made in this way. To construct these, a strong armature must be made, of wood covered with chicken wire – old palettes are very good for this purpose, or to build onto to extend the shape of the armature. This is covered with a skin of scrim soaked in runny plaster. More plaster is applied onto this, and then the plaster is built up until the final form is achieved. It can then be refined by using a surform and files. Walling plaster called Bonding coat can be used for such constructions.

On a smaller scale, plaster sculptures can be made over an armature of polystyrene, wire, wood or even screwed-up paper. The plaster is modelled in its cheesy state. Some artists mix straw with the plaster to achieve a very strong sculpture that will not break easily.

A portrait head can be made by directly modelling plaster over a head armature, in exactly the same way as for the clay head project (see Chapter 9). Features can be refined whilst the plaster is still wet; indeed the plaster can be soaked after drying out so it can be reworked.

Plaster can be coloured with oil paint diluted with linseed oil, or painted with acrylics. Plaster sculptures

---

MAKING PRESS MOULDS

Making a mould and producing a cast from that mould are both basic sculpting techniques. The simplest way to make a mould is the press mould technique, and it is a very good way of introducing casting. Both making the mould and the casting process are very straightforward, and because the clay is soft it allows easy separation of the mould from the cast. Making a press mould is also worthwhile as a means of familiarizing yourself with working with plaster; you will learn about its 'life', that is, how long it takes to go off.

Use a big block of clay, such as a new bag; this may seem extravagant, but you can re-use the clay for more press moulding or for modelling afterwards. Simply press objects – tools, fruit, in fact anything to hand – into the clay block to create negative spaces. You can even use parts of your body: push your finger in, or step onto the clay. These negative spaces can then be filled with plaster to create positives of the impression made.

It is important to wait until the plaster has completely set before removing it from the clay. You can check this by running your finger over the surface of the exposed plaster: if there is any moisture or slime present, the form is not ready to remove. It is a fun way to start experimenting with plaster, because you are never quite sure what you will end up with. If you are planning a composition using this method, you should remember that the cast will be reproduced the other way round from the clay impressions: what reads left to right in the clay impression will read right to left in the plaster cast.

Using this method, try pushing or cutting into the clay to create some big negative spaces, opening up the clay as much as possible. You can use implements such as a potato masher to push in to give interesting surface detail. You could also try incorporating found objects such as twigs or cloth; sheets of plastic will give a beautifully smooth finish to the plaster once it is cast. You will probably find that this process yields some very interesting forms; these may inspire you in making your sculpture, or they could even be cast into bronze. The process described here is exactly the same for the wax press mould projects.

---

are not really suitable for siting outdoors as rain will eventually cause the plaster to go soggy and deteriorate. You can, however, seal plaster by soaking it in linseed oil which will protect it and make it waterproof.

5

Working in Wax

Wax is the traditional material used to make sculptures to be cast into bronze, and it is undoubtedly the most direct medium because it can be worked in many ways: it can be modelled, used for constructing, and carved, as will be demonstrated in this chapter. But perhaps more importantly it is a material that, because of its inherent strength, can in most cases support itself so that no armature is necessary; it is also one that will last literally for centuries providing it is kept cool and out of direct sunlight. Indeed there is a surviving wax made by Michelangelo in the Victoria & Albert Museum in London.

Wax also has the distinct advantage of not cracking or having to be worked wet, as in the case of clay or plaster. It can also be worked using very little equipment and in a close confine, for instance on a board on your lap, or on a small table. It is less messy than working with clay or plaster, and can be controlled more easily.

TYPES OF WAX

What kind of wax should you use? The traditional form of wax used in ancient times, and most certainly during the Renaissance and since, is pure beeswax, although sculptors such as Michelangelo and Cellini formulated their own based on beeswax recipes. These days, however, the most commonly used wax for sculptors is microcrystalline wax, a synthetic wax produced as a byproduct of petroleum refining and so named because the crystals in its structure are smaller than those in natural waxes. There are over forty types of microcrystalline wax; advertisements by suppliers of the ones most commonly used by sculptors can be found in the publications listed at the back of this book.

There are cheaper alternatives to microcrystalline wax such as paraffin and earth wax, but these have not got the same properties and could prove very frustrating to use. Likewise, never attempt to use normal candle wax for modelling a sculpture. Pure beeswax is

---

TOOLS AND EQUIPMENT

The tools and equipment used for working in wax are basically straightforward and inexpensive.

Modelling tools You will need a small selection of steel modelling tools. I generally use one for nearly all operations; it has a double-ended blade, each blade being slim and flat and approximately ⅛in (4mm) wide.

Old dinner knives These are very useful for 'hot knifing'. Make sure they have some form of handle so you do not burn your fingers.

Craft knife An invaluable tool for cutting sheet wax and pieces of solid wax.

Old saucepan or double burner An old saucepan is fine to use for melting wax as long as you do not leave it unattended, and provided you keep a low heat. You can buy special wax melting pots from some suppliers, but these are expensive and tend to have a small capacity. Alternatively, a double burner is the correct method for melting wax, but these also tend to be quite expensive.

A heat source When working with wax a heat source is necessary for heating tools. For the hot-knifing technique you can use either a conventional gas or electric cooker; a more flexible arrangement is to use a small camping gas ring; this is ideal for melting wax as well. A normal candle held in a good holder is fine for heating small tools, and is what I use all the time for making my wax sculptures.

A soldering iron An electric soldering iron can be useful for softening wax and hot knifing. Irons can be bought with interchangeable heads.

---

very expensive and has no advantage over microcrystalline wax; in fact its consistency is similar to microcrystalline 'modelling wax': soft and rather gooey. Personally I favour the harder microcrystalline wax: when placed in hot water it softens nicely and can be modelled very easily by hand without having the same stickiness as the modelling wax, and can therefore be controlled better; it has the versatility to be cast, constructed or modelled.

Most waxes are sold in big slabs by weight in kilos and are normally neutral in colour. It is worth buying some wax dye which is sold in primary colour cakes to add to your molten wax. You will only need the smallest amount of dye to an average size saucepan: to approximately 1kg (2.2lb) of wax, use an amount of dye about the size of a pea.

MELTING WAX

Using molten wax is safe providing you follow a few simple rules. If you decide to buy a special wax melting pot then you need not worry about burning the wax as the heat is thermostatically controlled, and the molten wax will be ready to pour straightaway. However, if you decide to melt wax in an old saucepan, it is extremely important that you do so on a *low* heat, about setting no. 2 on an electric cooker. If the heat is too high there is a danger that the wax may combust into flames. If this happens, use a fire blanket, *not water*, to put out the fire; molten wax is like hot fat in that neither likes water, so never pour wax onto a surface where water has been spilt. When pouring into plaster moulds it is important that the plaster has been totally saturated with water; as long as there is no surface water left, it is safe to pour in the wax.

If the melted wax is allowed to become too hot you must not pour it because it will spit when it hits the moisture of a plaster mould. To test to see if it is the right temperature to pour, simply dip a paintbrush into it: if it hisses and burns the bristles, it is too hot.

When remelting a large quantity of wax that has set in a pot or saucepan, make a hole through the wax to the bottom of the pan; this releases pressure as it heats and melts. Otherwise pressure can build up, and – in the worst scenario – could cause the wax to explode.

It is also important to keep your wax as clean and dust-free as possible because any impurities in it may affect the quality of the final bronze.

CASTING SHEET WAX

Both sheets and rods of wax are very useful to have ready to hand for constructions and for modelling. To cast sheets of wax you can use either a marble slab or a sheet of glass, canvas or linen, or you can use a tailor-made plaster mould. An ideal thickness for sheet wax is ⅛–¼in (3–6mm).

A **marble slab** is ideal for casting sheets of wax; and if you cannot obtain marble, a **pane of glass** will suffice. Both need to be oiled lightly with any sort of cooking oil prior to use. The wax needs to be contained to the size you require; a strong clay wall or a frame made from slats of wood or metal can be used. Wood or metal should be sealed with clay around the outside edges; wood should be oiled inside as well.

A piece of **canvas** or **linen** can be used. These must be well soaked in water before use, but remember that hot wax does not like water so do not have puddles of water on your fabric, and do not pour the wax too hot.

A **plaster mould** is a very good way of casting sheet wax because it can be used over and over again, and is relatively simple to make.

To make a Plaster Mould for Casting Sheet Wax

First decide on a suitable size. This will depend on how much wax you can melt in one go, and how big you intend to make your sculpture; obviously moulds will need to be bigger if you are going to construct a very large sculpture. I would suggest a good size to be about 12in (30cm) square. Then find a piece of wood the right thickness – approximately ¼in (6mm) – and cut to the size required. There is no reason why you can't use two pieces of wood of different thicknesses in the same mould to enable you to cast two thicknesses at the same time, one thick and one thin.

However you decide to do it, it is extremely important that the edges are bevelled or chamfered so the wax can release itself from the plaster mould. Use sandpaper, or, even better, a sanding machine to achieve this. The bevelled edge needs to be gentle and even. I used sheets of balsa wood to make the mould in the photograph; it is very easy to sand.

Place your piece of wood with the wide side of bevel facing down on a sheet of polythene or clingfilm; oil the piece of wood with either cooking oil or washing-up liquid to act as a release agent. Make a clay wall around it, leaving a gap of about ½in (13mm) between the two; the wall will need to be about 2in (5cm) high, and it should be reinforced by placing thick pieces of wood around its outside. Nailed or glued together, these pieces will prevent the wall being broken and the wet plaster from spilling out when it is poured.

Mix up enough plaster to fill the mould. If you are using a 12in (30cm) square piece of wood, half a washing-up bowl full of plaster should be more than

Sheet wax mould. Two pieces of balsa wood have been placed in the middle of a strong wooden frame. Note the bevelled edges.

adequate. Pour the plaster in and leave it for about half an hour, or until it has gone off. If you use a light wood such as balsa you will need to hold the wood in place initially with two fingers while you pour the plaster.

Turn the mould over and remove the clay wall. The wood should release itself from the plaster leaving a recess for the wax to be poured in.

The Casting Process

Whenever you use your mould, the plaster should be totally saturated with water; it is advisable to soak the mould overnight before use to make sure of this. Before pouring the wax, make sure that the mould is on a flat surface, and that no surface water remains, otherwise spitting will occur. And remember, never pour the wax too hot, to avoid spitting.

Pour the wax slowly and evenly into the mould. After a few minutes you will see the wax become opaque as it cools, and it will release itself around the edges as it shrinks. If you wish to cool it quickly you can run cold water over the mould. You may, however, wish to use the wax sheet while it is still warm, especially if you want to bend the wax or model it; though make sure that it has set properly. Normally the middle of the sheet cools last; you can see if the wax in this area is still gooey.

CASTING RODS OF WAX

Wax rods are a very convenient form of wax to use in the modelling process. They are also an essential component in the bronze casting process where they are used for the 'runner' and 'riser' systems (see Chapter 16); a plaster mould for casting wax rods is thus normally known as a 'runner' mould. Here, however, we are going to use rods as a wax form. Obviously the advantage of making a plaster mould is that you can repeatedly cast many rods of an exact thickness and form.

If you wish to cast only a few rods you can do this more simply. Take a bed of clay or a new bag (the clay can be re-used). Cut a piece of dowelling of a suitable width to the length you want, and press it into the clay; the clay should not be too wet otherwise it will stick to the dowelling and spoil the impression. Remove it carefully to reveal the cavity to be filled with molten wax to make your wax rod. This is a simple press mould.

To make a Plaster Mould for Casting Wax Rods

To make your plaster mould, first of all roll out a large slab of clay with an even thickness of about 1½in (38mm) (see Chapter 3, the paragraph on rolling out under Slabbing Clay). The dimension of the slab will depend on the length of the pieces of dowelling used; I would suggest a slab approximately 26in (66cm) long by 10in (25cm) wide to be a good size.

For a mould of this size the pieces of dowel would need to be 24in (60cm) long, to leave a 2in (5cm) space at one end. Cut the dowel to length. You do not have to use the same thickness of dowelling in your mould. For instance, you could use one ½in (13mm) in diameter, one ¼in (6mm), and one ³⁄₁₆in (5mm). Press the dowels half-way into your bed of clay, leaving about a 1in (25mm) gap between each length and at the sides (see photo). The dowels will survive better if they are varnished and oiled prior to casting.

Build a thick, strong clay wall approx 1½in (38mm) high to surround the bed of clay; it can be reinforced with slats of wood to form a box, thus preventing any spillage when the plaster is poured.

You now need to make some location pins so that the two parts of the mould can be accurately fitted together when the wax is poured. Push the handle tip of a wooden spoon or similar implement into the clay to make distinct indentations. Make these location marks at intervals – say, every 3in (7.5cm) – along the sides and bottom of the clay bed.

Mix up enough plaster to fill the mould – approximately a washing-up bowl full – and pour it

Wax rod mould. The bottom half has just been cast into plaster; the wooden dowelling is still in place in the plaster. Note the location pins around the side. The plaster and the dowelling has been oiled with a release agent of washing-up liquid, and the top half is now ready to be poured.

while it is still runny. When the plaster has set you can remove the clay wall and turn the mould over; you should also remove the clay bed. Rebuild another clay wall of the same height and reinforce this with wood.

A release agent is necessary to prevent the two halves of the mould sticking together: mix some washing-up liquid with a small amount of water and brush this over the upper surface of the mould and the dowels left *in situ*. Then mix up another bowl of plaster and pour this over the first half of the mould. When it is set, remove the clay wall and carefully prise the mould apart; use a screwdriver to ease the join apart. Remove the dowels, and thus reveal your rod mould.

Making Keys

Before filling your mould with molten wax you will need to make some metal 'keys' for locking the two halves of the mould together; these can be made out of 6in (15cm) nails or a piece of ¼in (6mm) metal rod. Put the nail or metal into a strong vice and hammer one end over at a right angle. Then measure the thickness of your mould when it is closed, top and bottom. Turn the nail up the other way in the vice, and measure the thickness of the mould onto the nail, hammering the other end over to a right angle so that the nail becomes a 'U'-shape; this can then be tapped over the mould to form a clamp, closing it securely top and bottom on both sides.

The plaster must be totally saturated with water before the molten wax is poured in; if necessary, soak the two halves overnight. With the keys securely in place the mould can now be filled with wax. If the mould has been well made there should be little or no flushing on the wax rods.

MODELLING WITH WAX

When making a wax sculpture you will probably find that at some stage you will employ all three of the main techniques for working in wax: modelling, construction and carving. This is certainly true of the two wax projects in this book. Remember, whatever technique you use to work with wax, you must always try to keep the thickness of the sculpture to a maximum of about 1in (25mm). A large wax sculpture would need to be cast hollow and would therefore need a core (see below, Predetermined Cores). It is possible to cast solid bronzes several inches thick, but there is a risk of

'shrinkage'; this will show up as holes left in the surface of the bronze, and is a result of the way thick bronze cools down. It can be counteracted by using thicker risers or multiple risers on the wax, although a foundry will advise you on this when you take along your final wax to show them.

Modelling with wax is very similar to modelling with clay, except you will need to warm some types of wax first to be able to manipulate it easily with your hands (this only applies to the harder forms of microcrystalline wax). You can do this by cutting small lumps and putting these into hot – not boiling – water for a few minutes. If you leave the wax in the water for too long you will notice the colour changing as it goes to a gooey stage. If this happens you can simply cool it by placing it in cold water, then soften it again in hot water.

When the wax is sufficiently soft you will be able to squeeze it into shape with your fingers easily without it tearing or going gooey. However, you may need to practise until you learn how hot the water should be and how long to leave the wax. If you are using modelling wax you will not need to soften it in water as the heat from your hand should soften it sufficiently to use.

You will find that you can shape a form such as an arm or leg just by squeezing the wax into shape. To model a larger form such as a head or body you will need to keep adding small lumps of wax to build up the structure, in exactly the same way as in clay modelling.

Modelling wax is very sensitive to detail; your fingerprints may be picked up on the surface.

CARVING WAX

It is possible to achieve a wax sculpture purely by carving from a single block of wax using a craft knife; it is easy to cut pieces away and shape forms. You can also use a hot knife to carve forms and detail. A certain amount of carving is used in the modelling process, but you may find that you would like to tackle a carving purely by removing wax, and not adding it. The final bronze may then have the look of a wood carving.

Start by cutting off a block of wax from a large slab. It is useful to do a drawing of this block onto paper, and draw onto it the form you wish to create (see diagram). By shading in the areas in the block that are around your form you will know which areas to remove from the wax block. You can mark these points on the wax,

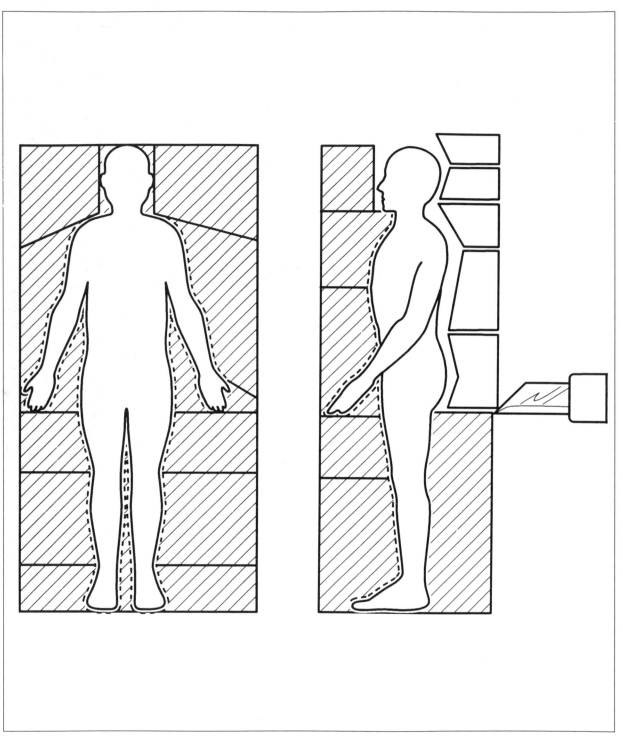

Blocking out a figure from a block of wax. Draw the figure onto the wax from all sides, scribe in cutting lines, and then cut away the wax to get to the form.

and then cut away the area. You will have to think about what is happening from the other sides as well, so try to draw the form in the block from all sides to assist you in what to cut away.

CONSTRUCTION

Constructing a sculpture in wax normally takes the form of cutting sheet wax into shapes, possibly bending these, and then welding them together along the edges using a hot knife. If you were making a large sculpture out of sheet wax it would be advisable to reinforce the inside with strips of wax to stop it distorting. You can use a combination of modelled and carved forms to construct a sculpture, also incorporating sheets and rods of wax.

In the example shown in the photograph the artist has used sheets of wax to achieve a columnar form. She used warm sheets and cut them to size before rolling them over into a hollow cylinder. The joins down the sides of these columns were welded up using a hot knife. The form on top of the sculpture was a real twig which was welded on with a hot knife; it then

Bronze sculpture by Barbara Cheney. An example of a constructed wax sculpture; 14 × 7 × 6in (35.5 × 18 × 15cm).

HOT AND WARM KNIVES

You can either smear the wax over with your fingers to achieve a continuous surface, or use a hot knife. This can be any form of steel modelling tool: an old kitchen knife, a teaspoon or even an old hacksaw blade. To heat your knife you can use either an electric hot plate or a gas ring such as a gas cooker or camping stove.

You will soon learn the difference between a hot and a warm knife: a hot knife will fizz as you apply it to the wax and can melt away quite large amounts; if used accurately and quickly you can obtain very strong welds between surfaces of wax. This is crucial in the investment process when the runners and risers are attached to a sculpture to be cast into bronze (see Chapter 16). A warm knife will be very useful in the modelling process, especially when creating fine detail, as it can be used to mark the surface of a sculpture and to carve in features (see photo).

The beauty of modelling with wax is that you can build up quite a large form by sticking many pieces of wax together and the sculpture will support itself, providing the wax is kept cool and is not too thin.

See Chapter 15 for a detailed project on how to model wax.

---

burnt out with the wax (see Chapter 6, the section Using Found Objects). This bronze is a good example of constructed wax, the main forms appearing soft in contrast with the more spiky forms which are attached to them. These spiky forms are in fact defects in the casting process, but ones which the artist decided to leave on to add to the organic feel of the piece. Part of the riser system has also been left on to add to the overall composition.

Another interesting technique is to make clay

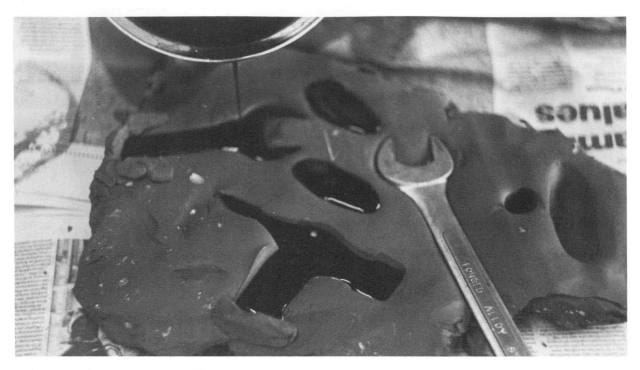

Pouring wax into a clay press mould.

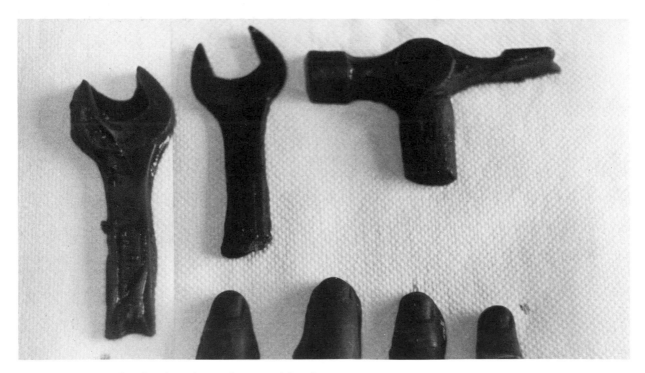

The wax positives after they have been taken out of the clay.

Hot knifing forms together. Note the pool of wax at the end of the knife. You have to hold one form in one hand and the hot knife in the other, and while the knife is still very hot place it on the surface of one form and quickly bring the other form down onto it; then remove the knife as quickly as possible to obtain a very strong weld.

PREDETERMINED CORES

All large bronze sculptures have to be cast hollow, so a core is made and inserted into the hollow wax. This is made of grog (the ceramic material used for investing waxes) and held in place with core pins. (This process is described in detail in Chapter 16.)

Predetermined cores have been used worldwide since the early Egyptians, including throughout Asia, India, Africa and Europe. A clay core was modelled into the shape of the intended sculpture, often over pieces of wire which would act as core pins (pins to hold the core in place). The wax was then modelled to the required thickness over the core.

When making a large wax directly, a predetermined core would be used. For example, with an under-life-size head or a large bulbous form of wax, a core made of polystyrene, paper or wood could be used: you would need to cut the core material to the rough intended form and then model the wax over this, provided the form were not too big. Once the sculpture is made, the wax can be cut in half and the material removed; the wax can then be stuck back together and filled with the grog core. The foundry would cut a small hole in the wax to facilitate this process, and the hole or cap would be welded back in place when the casting process has been completed. A structure the size of a life-size head would need a rubber mould to support it while the core was poured in.

You may find this a useful way of working when making a more bulbous sculptural form, as it allows you to work on quite a large scale without having to employ the expense of a rubber mould.

impressions using various objects such as steel tools, pieces of wood and metal, hard textured materials and even parts of your own body (see Chapter 4, Making Press Moulds). Molten wax can then be cast in and left to cool. Exciting assemblages can be made by juxtaposing these pieces, joining them together using the hot knife technique (see photograph).

To add to the variety of form you can also use real found objects with the wax to make inventive original sculptures. The project in Chapter 18 describes such a process in detail.

6

Methods of Construction

Constructed sculpture (also known as 'assemblage' or 'collage') is the process of using materials that cannot be modelled, but can be shaped and then stuck together. Picasso was one of the first artists to be credited with inventing this kind of art form: with *Musical Instrument* (1914) and *Still Life with Fringe* (1914), Picasso demonstrated for the first time that a sculpture could be made out of parts just as a craftsperson might make a table or chair. Picasso also demonstrated a direct response to the materials he used, and a straightforward crudity of construction unalloyed by preconceptions of style or taste.

This type of sculpture also became known as 'object' art: that is, the constructions used still-life objects as their subject matter and were actually made out of the same materials and in the same way as the real, made objects. Thus *Still Life with Fringe* depicts a shelf or table with a real pom-pom material fringe surrounding it; the rest of the construction is made with wood depicting a knife and a slice of bread with sliced meat on it, all made out of painted wood. Picasso went on to form a collaboration with Julio Gonzalez, and over a period of four years the two men made a series of constructed steel sculptures that have influenced the Russian constructivists, the American sculptor David Smith, and successive generations of constructed steel sculptors in Britain and abroad.

Picasso's original constructions also include the use of so-called 'found objects', ready-made or natural objects used whole or in part. The fringe in his still life, as already mentioned, is a found, or ready-made object, and other famous examples include the toy car which becomes a baboon's head in *Monkey with Young* (1951), and the *Bull's Head* (1942) which uses the simple juxtaposition of a bicycle saddle as the head and the handlebars as the horns. The actual process of construction that Picasso used has changed little: finding interesting materials and everyday objects, and with a bit of inventiveness and playfulness turning them into works of art.

The object of this chapter is to demonstrate the use of materials and found objects that can be successfully

---

MATERIALS

A variety of materials can be used for constructive processes: wood, paper, cardboard, light wooden boxes, packaging materials, fabric, string and papier mâché; in fact anything that is light enough to burn out in the kiln. Moreover, there are many materials that no sculptor has ever attempted to cast into bronze, and you may well think of a new way to make a sculpture using some form of everyday material. Many unusual materials have been cast into bronze, including rolled-up newspaper, bandages, part-baked pastry dough, and even a real teddy bear. These materials all burn out successfully in the kiln, although they may prove to be quite difficult to sculpt with. As regards found objects, one aspect which seems to be quite in vogue at present is the process of casting dead animals into bronze. Indeed, some artists have cast dead fish, hedgehogs and birds' wings; these objects all burn out in the kiln, and the bronze picks up every single bit of detail. The results look very impressive, although the exercise may be more technical than creative.

---

burnt out in the kiln directly: that is, a sculpture that can be made out of a variety of materials and combinations of these materials. The wax runner and riser system as described in Chapter 16 can be directly attached to a sculpture made out of wood, paper, card or whatever, and then invested, and the materials all burnt away to leave the exact negative shape of them inside the mould. This does mean, however, that the sculpture made of these materials has to be sturdy enough to support the weight of the runner and riser system. It also has to be able to withstand the wetness of the first coat of plaster in the investment process. Nevertheless it is possible to strengthen any weak parts of a sculpture, or fragile, delicate materials by brushing onto them some molten wax; thus a piece of paper or cloth can be thickened slightly and strengthened by wax without taking away any surface detail you may wish to capture. Remember to apply the wax with a paintbrush, letting it just fall off the brush.

DEMONSTRATION: USING WOOD, PAPER AND CARDBOARD

Making a sculpture out of paper, wood and cardboard offers a cheap, direct alternative to the other techniques talked about in this book. The equipment needed to make such a sculpture is minimal: a craft knife and some glue, though in addition a glue gun is an extremely useful piece of equipment, as the glue is very quick-drying and strong. I have selected a range of scrap and inexpensive materials including card, corrugated cardboard, balsa wood, dowelling, watercolour paper and an old cheese box. Constructing with such materials is an exciting business, and will put your inventiveness to the test. You can transform an everyday object into something quite different, and once cast into bronze a constructed object will have a completely different quality to it. Bronze picks up every subtlety of form, so any indented writing or texture will be perfectly reproduced.

This construction was made with just a craft knife and a glue gun, and the sculpture was made in a direct way, with no prior idea of what the final form would be. Obviously you could use the materials to construct a specific object such as a boat, a car or a cubist-type person. The cheese box was explored first and cut in half. The curved piece on top of the sculpture was made next, and related or juxtaposed to the box in different positions. The rest of the sculpture evolved around these two pieces, making use of the two types of paper, rolling one up and using the other in a more level, flat way. The round circles were made by cutting unevenly through a piece of thick dowelling. The whole sculpture was made using a process of cutting, shaping and sticking pieces together, trying this piece here, turning it upside down and trying it again there, composing an arrangement of parts that felt satisfactory, similar to a piece of music that has a base line, the main 'body' and the melody jumping around this, with another instrument playing a few quirky frills.

Constructed sculpture lends itself to this kind of orchestral arrangement because such structures can be very open and rhythmic. Have a look at Anthony Caro's *Midnight Gap* (1976), a constructed steel sculpture with one large plate of steel obliquely raised off the ground, and a whole circle of different shaped and sized pieces of steel dancing around, set above this plate. When making a sculpture in this way you can take a risk and go quite wild with your judgements, playing around with various materials and forms, trying to achieve a variety of parts to make a very visually interesting sculpture.

A selection of materials including balsa wood, dowelling, paper, corrugated paper and an old cheese box. Note the glue gun.

The final constructed sculpture; 11 × 7 × 2in (28 × 18 × 5cm).

PAPIER MÂCHÉ

Using papier mâché offers a cheap, clean and easy-to-use material that is light but durable; it can be made using any old newspapers or old paper and light fabric material. All you need do is cut the paper into strips and soak these in wallpaper glue or size, the only cautionary advice being not to make the layers too thick or they will take an age to dry out. The glued strips can be placed over an armature of either chicken wire, cardboard or polystyrene. With a large form, life size for example, an armature of wood may need to be used, over which chicken wire or expanded metal sheeting can be attached.

To make a papier mâché head, a rubber balloon can be used, the strips of paper being built up consecutively over the inflated balloon. Features could be modelled onto this with paper pulp. Papier mâché pulp has a soft and plastic consistency and can be used in a similar way to clay. It is made by cutting newspaper into strips about 2 × 6in (5 × 15cm) and soaking these in wallpaper paste overnight; they are then briskly stirred – an electric mixer comes into its own here – until the paper has broken down. Such a head could be cast into bronze: a hole would be cut into the dried paper and a core poured in, and the papier mâché would burn out directly in the kiln. The effect of the paper strips for such a head could look very interesting in bronze.

Small figures or animals could also be modelled out of papier mâché pulp.

Bronze relief by Barbara Cheney. This relief was made from wax, organic found objects and corrugated cardboard. The form on the right was a whole piece of hollow bark. All the objects were secured to a wax sheet by molten wax poured and–brushed on. At the top left is a bird's feather (see detail below); 16 × 12in (40.5 × 30.5cm).

USING FOUND OBJECTS

Found objects, as already mentioned, can be ready-made or natural objects. A ready-made object is literally one that has been ready-made for another function or purpose, such as the teddy bear described earlier, for example, which was cast directly into bronze, a core having been placed inside after the stuffing had been removed. This was exploiting a ready-made object solely on its own; however, it could have been incorporated into a construction involving other materials.

Natural objects such as dried flowers, fruit, seeds, cones, leaves, sticks – indeed, any organic form that would successfully burn out in the kiln can be used, though it should be said that there is always a risk with using found objects as opposed to pure wax that something could go wrong either in the kiln burn-out or in

the bronze-pour. If a foundry felt that they could successfully cast a delicate, technically more difficult sculpture, then obviously they will do so. The investment of found objects is far more tricky than with wax. The blackberries in the sculpture shown in Chapter 8 were incredibly difficult to cast because they had to be done so from frozen. The first coat of plaster takes some time to establish on the form being cast, and the plaster was only just about thickening up by the time the berries were starting to thaw out and go soft.

The bronze bird (see photograph) was made up of nearly all found objects: the base was a mushroom cap, picked and immediately dipped in cool molten wax to preserve it. The legs were made out of dried artichoke stems; I simply pushed them into the mushroom cap and sealed around them with wax. This bottom part of the sculpture was cast separately, but in the same mould as the other part of the bird, and the two parts were welded together later. This was because the mushroom was delicate, and would not have been able to withstand the weight of the top part of the sculpture. This is often the case when using found objects and thin wax: forms may have to be welded together on completion of casting. The wings were made from a piece of corrugated cardboard. The breast was half a dried artichoke head. The back and neck were modelled wax, and the head a dried poppy with wax eyes. The chest was glued to the cardboard, and the back was hot-knifed to the cardboard. The head was pinned into the wax using part of a poppy stem.

Chapter 18 gives a detailed description of a sculpture made out of wax and found objects. Found objects can also be incorporated into mixed media constructions, or used in conjunction with clay, plaster and wax.

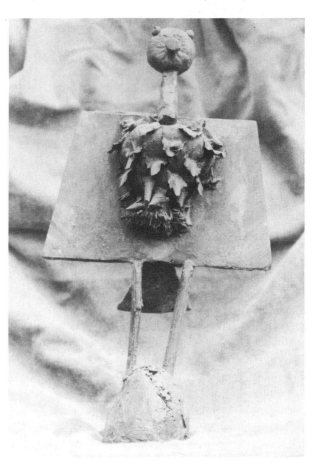

Bronze bird by Guy Thomas (front view);
12 × 6½ × 2½in
(30.5 × 16.5 × 6.5cm).

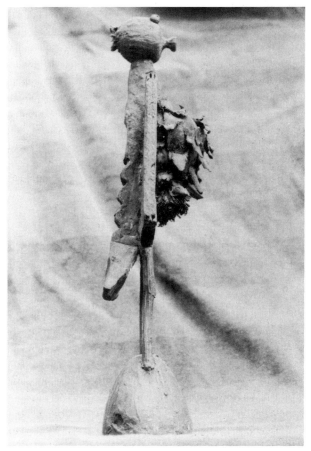

Bronze bird (side view).

7

Methods of Life Casting

'Life casting' literally means taking a life cast from a part of the human body. We have probably all seen, in museums or in reproduction, the death masks: a cast taken from a famous person soon after he or she has passed away. Life casting uses exactly the same techniques and has been practised by sculptors since the great ancient Mediterranean civilizations. Casts from life have been valuable to sculptors as aids in studying the human figure, especially as a reference for modellers and carvers alike. When Rodin first exhibited his sculpture called the *Age of Bronze* (1875), he was actually accused of casting it from life because the modelling was so life-like.

```
┌─────────── A PERMANENT RECORD ───────────┐
│ Life casts are used frequently in the performing arts, for │
│ films and in the theatre. The face cast which is turned │
│ into bronze can look very impressive and serves as a │
│ record for somebody's life. Likewise a face cast of a │
│ child would have enormous sentimental value later on, │
│ capturing a frozen moment in the same way as a │
│ photograph. Sculpturally one could exploit such │
│ methods, incorporating life casts with a modelled fig- │
│ ure or making an assemblage of parts to create a dis- │
│ turbing but fascinating object. │
└───────────────────────────────────────────┘
```

In the past, life casts were made in gesso, a type of plaster, and whiting (mineral and animal glue) made from gypsum. Parts of the body such as ears and whole legs were cast, and even women's busts – quite a common practice amongst art students today.

Nowadays there are four main materials used for life casting: plaster, plaster bandage, foam latex and alginate.

USING PLASTER

Plaster is the traditional material used for taking life casts. You may wish to try a plaster cast instead of using the rubber materials, although it could prove to be slightly more uncomfortable to experience than latex or alginate because of the time it takes to go off. With any life casting, the sitter has to be able to keep that part of his or her body perfectly still. In fact as it goes off plaster will inevitably keep the limb or part of the body still because of its rigidity; obviously the sitter would have to endure carrying the plaster cast until it has gone through its warm phase of setting. The latter can often be a pleasant experience as the warmth of the plaster radiates through that part of the body.

The most important thing to remember when using plaster for life casts is to apply a release agent to allow easy removal of the plaster from the body. Vaseline is essential for this process. Even so, for people with hairy limbs a life cast can be a very painful experience, as the hairs will undoubtedly get trapped in the wet plaster no matter how much vaseline you use. Shaving the area may well be worthwhile as a precaution against this happening. When casting a large area such as a leg, use scrim to reinforce the plaster.

USING PLASTER BANDAGE

Plaster bandage is literally bandage that has been impregnated with plaster. It comes under the trade name of 'Gypsona' or 'mod-roc' and is the same as that used in hospitals to set broken bones. Once the bandage is dipped into water the plaster becomes activated and will set within three to four minutes. It can be used to produce a thin, strong, light cast.

When using plaster bandage, the area needs to be greased with vaseline first. Because plaster is activated by water do not let the bandage come in contact with it until you are making your cast. Build up only to a thin layer – approximately two or three layers – as it is very strong when it dries. It can be removed whole or snipped away in sections to be rejoined off the body by bandaging the pieces together again. The plaster shell-like mould can then be used to cast wax, plaster, cement or fibreglass.

USING ALGINATE

Alginate-hydrogum is similar to latex in its final consistency, but comes as a dust-free powder. It is the same material used by dentists for taking impressions of teeth which are then cast into dental plaster.

Alginate is very easy to use and very quick drying. Its working life – that is, the time it takes to set – is approximately two minutes, which means you have to work very fast and accurately. As well as life casting, alginate could also be used for taking a mould from other materials such as clay, plaster, wood and metal. Its only disadvantage is that it is not very durable, so you will only be able to take a few casts from the mould and these have to be taken as soon as possible. However, you can cast wax, plaster or even fibreglass from the moulds. Alginate could be very useful as an alternative to a rubber mould-making material as it is direct to work with. This would only be possible on a smaller sculpture – a life-size head would prove too big to cast in one go, although you might try it in several parts.

DEMONSTRATION: USING ALGINATE TO CAST A HAND

This is a process similar to using plaster to cast a hand, but it is much quicker and easier. First of all, make sure that your sitter is made comfortable; he or she will have to keep his/her hand still for ten to fifteen minutes at a time. Grease the hand with vaseline; the alginate will not trap hairs so shaving should not be necessary. Push the hand into a bed of clay, making sure the fingers are half buried in the clay all the way around. You will need to smooth the clay around the fingers to make sure that there are no gaps or dips in the clay for the alginate to fill into. With the hand firmly embedded and kept still you are now ready to mix the alginate.

This should be an equal mix: the same quantity of alginate to water. You will have to work quickly because you will only have about 2½ minutes to cover the hand evenly. The thickness of the alginate should be a minimum of ⅜in (9mm) all over, covering the hand surface as evenly as possible.

The alginate mould will need a support case. Working swiftly, as soon as the alginate has set, mix up enough plaster to cover it; the thickness of the plaster case should be about 1–1½in (25–38mm) and this can be reinforced with scrim to make it stronger. Wait for the plaster to go off before lifting the hand upwards away from the bed of clay. There may be some degree of suction from the clay, and so you may need to push the clay down in order to release the hand. Turn the hand over with the alginate and plaster case still on it. The face of the plaster case needs to have some holes scribed in it to act as location pins; this is exactly the same process as marking location pins in a clay wall (see Chapter 11). The plaster, although it has set, will still be soft enough to be able to mark the holes with the tip of a wooden spoon handle or similar implement. This is to ensure that the two plaster cases of the hand locate correctly.

Regrease the exposed part of the hand with vaseline. Brush a concentrated solution of washing-up liquid and water on the plaster face surrounding the alginate and on the surface of the alginate to prevent it sticking to itself. Still working quickly, mix up more alginate, apply it, and make a plaster casing for this half.

When the second half is complete, use a screwdriver with water sprayed into the join to prise the two halves apart: the hand can now be released. From this alginate mould you can either cast a plaster or a wax hand; for the purposes of a bronze we require wax. Molten wax can be painted into the two halves separately – remember always to let the wax just fall off the brush. Build up the thickness of the wax to about ¼in (6mm) and then join the two halves of the mould back together, sealing the outside crack with either wax or scrim soaked in plaster. Pour some molten wax into the mould to seal the two wax halves together; and try to guide the wax around where the crack is inside the mould. Carefully remove one half of the plaster casing, and then remove the alginate from the wax. Repeat this for the other half: plaster case first, then alginate, to ensure that you do not damage either the wax or the alginate mould.

The wax hand can now be finished by working on the seam line. You may need to employ the use of a warm modelling tool to touch in areas or to fill any cavities with wax. The extent of detail on the wax hand is amazing; every wrinkle of skin and line on the palm can clearly be seen.

To tackle a face mask you follow the same procedure, although obviously this is cast in one go. The important thing about taking a face cast is to let your sitter breathe: this is usually accomplished by putting straws up their nose. A plaster case should also be made to support the shape of the alginate mould.

The hand is in place in the clay bed and the alginate has just been cast over the top half of the hand.

The plaster case has been applied over the alginate with the hand still being kept perfectly still in the clay bed.

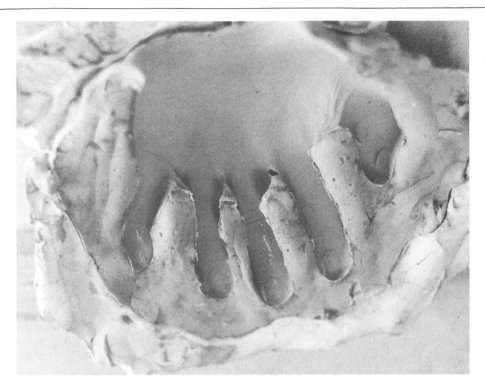

The alginate mould with plaster case supporting it.

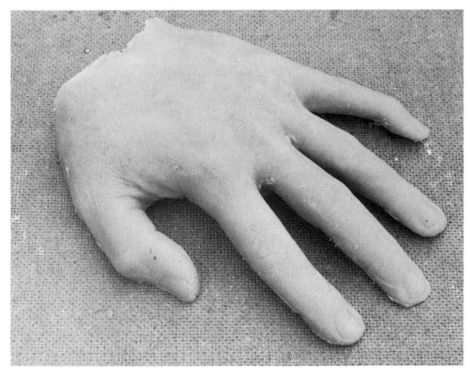

The final wax hand. Notice how life-like it is!

8

What Makes a Good Sculpture?

What makes a good sculpture? This is a difficult question to answer, but one worth considering as you become increasingly involved in this activity. It is important to consider sculpture generally because it will inform and help you in your own creations. It is a question that can perhaps be answered by considering the many conditions that exist in making up a sculpture. We can look at the many different types of sculpture, the variety of materials used, in sculpture both ancient and modern, Western and non-Western. You will undoubtedly explore this question for yourself by looking at as much sculpture as possible in museums and art galleries. And as you become more involved in sculpting activity, you may find a hidden part of yourself emerging, a seventh sense beginning to come into focus.

As time passes and your technical skills and your knowledge of form increases, so too will your attitudes change. You may find your tastes changing, that you discover new artists whose work feels 'right' for you; and in turn one sort of artist may be replaced in your esteem by another – you may find that you start off as a figurative artist but increasingly turn towards abstraction. You may wish to have a finger in many pies, tasting a variety of flavours. The more sculpture you make, the more you will take techniques for granted, and in your growing confidence will start breaking rules and exploring new territory.

At first you may feel inhibited about making a mess with your work, worrying about the correct procedure, about following the instructions carefully, thinking that there are set formulas to follow in making a sculpture. The art of making sculpture does rely upon many techniques, much more so than painting, but there is essentially no such thing as *a technique* because anything you do to affect a piece of material is a technique. There are still many to be invented, perhaps some by you as you have accidents with your work and find new ways of doing things.

As you tackle some of the projects in this book you will discover your own creative powers as they come into play. You will invent your own sculptures, and as you do so you may find yourself saying, 'Well, I like my sculpture – but is it a good one?' To answer that you will have to look into the sculptural mirror and often compare your work to that of the great masters.

Try to forget your inhibitions and have confidence in your own ability to make a wonderfully inventive sculpture. Making sculpture is a natural activity not bound by rules, so relax and enjoy yourself.

THE TRADITION OF SCULPTURE

To help us find our feet in the world of sculpture we can look at its long tradition, a tradition which is ongoing, new periods being invented and reinvented. You may wish to consider where you fit in, what period you would like to join; perhaps you feel you have to be modern and join this year's fashion parade.

I would like to hypothesize that formally sculpture has not changed since the earliest times, indeed many of the materials have not changed, either: sculptors today still carve stone, and they still have their work cast into bronze. The newest sculpture is not really new at all.

Installations were part of Neolithic religion and culture. The ancient Egyptians and many other cultures pickled dead animals – and indeed themselves – in what became a theatrical spectacle. The world revolves around, and so too do artistic ideas and modes of expression. What we can add to this tradition is our own personal interpretation which is unique.

If you look closely at the tradition of sculpture from all cultures worldwide you will notice the same themes emerging: mythology, birth, death, the depiction of religious leaders. There are many similarities in the expression of decoration and form. We have the advantage of making our own recipe for a sculpture using as many ingredients from this tradition as we like. Ancient, African and pre-Columbian influences can be detected in the work of Eduardo Paolozzi, for example.

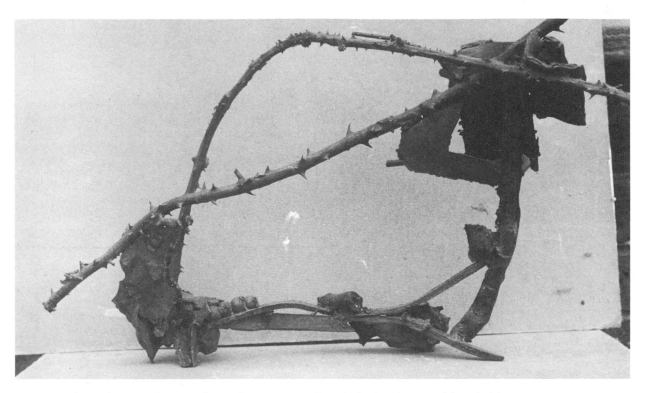

Bronze sculpture by Guy Thomas. This sculpture was made entirely out of wax and found objects, the found objects being the bramble stems and the rosehips. This sculpture is of a landscape format, the subject matter being the landscape itself; 16 × 25 × 6in (40.5 × 63.5 × 15cm).

THE NATURE OF SCULPTURE

As you start making sculpture you will begin to appreciate the differences inherent in this activity as opposed to other art forms. You will be dealing with real physical materials. This is quite unlike painting which, despite its sensual qualities, relies essentially on an illusionistic art form. With sculpture you are using solid materials, building in a three-dimensional space. You have total freedom: you can make your sculpture as big or small as you like, as thin or bulbous as you like. You have the freedom to build something that can stand in your space without the need of a base.

This physicality of sculpture as a three-dimensional form tends to challenge and can intimidate some people, because a sculpture is in a real space, your space. This imbues it with a presence; it is as if some sculptures have an existential being of their own, perhaps even a soul. In other cultures sculpture is made for religious reasons and not as art. Sculptures are used

as containers to trap spirits: once the spirit 'enters' a work it becomes trapped, and is appeased by ceremoniously making a sacrifice of a goat or a chicken, its blood often poured over the sculpture to give it a life force. In our culture we tend to place sculptures in galleries or on a special site outside where we can visit them and hold them in high esteem or even with contention.

Once made, a sculpture will lend itself either to a monolithic, upright, columnar format or to a horizontal, landscape format. A figurative sculpture can lend itself to both formats. Think of the great *Balzac* sculpture by Rodin: it would be almost impossible to get a more monolithic sculpture than this. And if you consider some of the Henry Moore reclining figure series, those figures become almost landscapes. Look at the sculpture in this photograph, and you can see that I responded to a landscape format – indeed the subject matter is landscape itself, using what were real brambles and rosehips with wax to create an organic sculpture. The brambles and rosehips were burnt out with the wax in the kiln (see Chapter 6).

The next time you visit an art gallery, have a look at the sculpture and see what kind of format they fill. When you come to make your own sculpture you will find yourself responding to either the monolithic or the landscape format.

COMPOSITION

Your response to either format is a natural, instinctive one, and you guide this response intuitively. It is the same response that orders your life, how you arrange objects around the house, how you arrange your meals on a plate, in what order you get undressed, whether you follow the same patterns every day. This sense of order can be called composition, a structural arrangement, the art of putting things together to form a whole. This has a direct link with making sculpture, affecting how you arrange one part to another, juxtaposing, making decisions, until you arrive at a final satisfying form. This sense of composition applies to all sculptural practices, whether it be carving, modelling, construction or casting: we all have this ability to make decisions to compose our daily lives and to compose sculpture.

As you start to look closely at the work of other sculptors you will notice composition, and you will ask yourself all sorts of questions: 'Why have they put that there? What an odd shape. That knee bends backwards instead of forwards, that doesn't make sense'. Often you will find some of the arrangements of certain sculptures very uncomfortable to look at, and this will tend to put you off the work of certain artists. In time you may find that, due to your own sculptural practice and experience, your attitude changes: 'I like that form after all, it makes sense to me now. The composition of that sculpture is a pleasing one. It challenged me at first, but now I really do like it.'

This process will tune your own sense of composition; through looking and making you will find your ideas changing. The opportunity to make a sculpture gives us the chance to break our own rules of composition. This may feel extremely uncomfortable at first; but . . . 'What would happen if I placed this piece of wood over there instead of here, where it feels right to my sense of composition? What would happen if I modelled this feature in a completely different style?'

CREATIVITY

People who take up art in their later years are often quite surprised to discover that they can be, and are, creative. This problem lies really with our upbringing and schooling. As we grow older we are conditioned to be serious, and are educated to go out to work and earn our living. The child we were disappears – and with it often disappears our creativity, too. This process happens very early on, unfortunately. As children we drew and painted as naturally as we ate, we played with toys and empty cardboard boxes, arranging objects: we were being creative. Taking up art again is rekindling that lost part of ourselves, finding a way of expressing ourselves in a powerful medium that can bring reason back into our lives.

Expression

If we can get in touch with what we feel inside, our hidden emotions and day-to-day frustrations, these feelings can be expressed through art consciously and

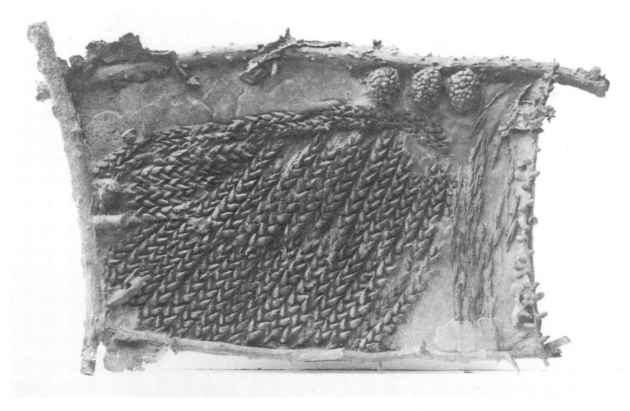

Bronze relief by Barbara Cheney. This demonstrates the creative use of organic found objects. These objects were placed on a wax sheet, and molten wax was poured and painted on and around the objects to secure them in position. This is a simple technique, but the artist has demonstrated a strong sense of composition and consideration about the placing of the objects. She has clearly thought about texture, and part of the expression has been captured in the areas of flashing that have been left on the final bronze, which can be seen clearly on the left of the relief top and bottom. The found objects were burnt out directly with the wax in the kiln; 10 × 16½in (25 × 42cm).

unconsciously; and form can be given to our expressions. The actual process of making art is a non-verbal one: it links what we see with what we feel, it is a bridge between our unconscious and our conscious minds, between imagination and reality. When you start looking at sculpture from different times and different cultures, and when you consider modern minimalism (Carl André's bricks, for example), you will be able to detect the extent of expression – which sculptures are born from imagination and which from conscious reasoning; which are the most interesting to you, and how do these compare to your own. Expression will influence your work, the way you feel will undoubtedly show in your sculpture. Some sculpture appears so imbued with expression that it seems to speak to you: it is vital, full of energy, pregnant with form.

Working Quickly

How can we nurture such sculpture from ourselves? How can we get in touch with our creativity without intellectualizing it? One good way is to make sculpture very quickly, even actually to set yourself a time limit. If you were to allow yourself, say, half an hour to model a head or construct a sculpture in wax, you would have to work very, very quickly. But this is a good exercise; it is a great way to warm up before going on to tackle longer, more considered sculpture. It teaches you to make decisions fast, without thinking too long about them, and forcing yourself to work this quickly brings out your creative potential and your sense of true expression.

It is quite difficult to work in this way to begin with, but if you can really push yourself you will find it a liberating experience. Your sculpture will be much looser, and it will have a fresher quality to it. You may find it useful to have somebody timing you, pushing you to work more quickly should you dare to stand back and consider for a few moments.

It is possible to make a modelled head in half an hour. You will have to work frantically, slapping clay on as quickly as possible, but it is well worth trying. It is a good exercise to remember, and will loosen up your responses and your sense of drawing.

DRAWING

Drawing is as important to sculptors as it is to painters. Painting is generally considered to be the art of drawing with paint, and likewise sculpture can be considered the art of drawing with form. If you ever listen to sculptors talking about sculpture you will often hear them refer to how well a form has been drawn. We have all drawn as children, and we can still draw, although most people consider that they cannot. But

Bronze sculpture by Guy Thomas. In this work I took advantage of the curved line of the bramble stem, incorporated as part of the drawing; the final piece demonstrates the rhythm of line moving around, changing form as it does so. At the top right I have left part of a riser on the final bronze to add to the sense of composition and to echo the short fat stub at the extreme left of the sculpture; 11 × 19 × 6in (28 × 48 × 15cm).

good drawing has to be worked at – like most things, it has to be nurtured from within us. The trap that many people fall into with drawing is to learn a specific technique and to stick with it, never changing this style; thus essentially every time they draw using such a technique they are drawing the same picture over and over. However, drawing is not about technique but about seeing and feeling, and the more you look, the more you see. Thus the more drawing you do, the more information you are gathering about form, space and composition; and so you will begin to understand form and composition until it becomes second nature to you.

Sometimes an artist will make a work that seems to be badly drawn, roughly made, random in its composition. Often these works are actually brilliantly drawn, although we are somehow led to believe that they are naive. It might be a worthwhile exercise to attempt to do a very good drawing, spending a long time on it, considering it very carefully, observing very closely and intently; and then to do a quicker, more casual and random drawing, even to attempt a really bad one.

PRACTISING YOUR DRAWING

If you can draw a bit every day you will soon improve your drawing sense, and this will directly influence your sculpture making, improving the quality of your forms in whatever material you happen to be working with. Along with more considered drawings, try some very quick ones, timing yourself to half a minute, one minute, five minutes and so on. You could choose anything for your subject: landscape, still life, your family, or even enrol for life drawing classes. Working from the figure is extremely beneficial, both for improving your drawing abilities and also to feed into your sculpture making. Even if you are an abstract artist, the naked body provides a real challenge to draw and a wealth of form, shape, line and rhythm.

Then compare them, and you may be quite surprised to find that the 'bad' drawing has a much more lively, rhythmic quality about it. Some of the mark making may be quite different from your usual style, and it may even appear messy to look at, but it could turn out actually to be a better drawing than the other one. Pick out the good qualities, the ones you like, and apply these to your sculpture. Drawing is as much a risk-taking activity as sculpture can be. Remember, play is the root of creativity. If you look at the history of sculpture, in many of the greatest examples the artists have all demonstrated different qualities of drawing.

TASTE

What makes a good sculpture is a combination of all the things talked about in this chapter, plus many more. Also, what makes a good sculpture for me may not be the same for you. I nearly always rely upon my gut reaction when first looking at sculpture. If we were all given the same size block of wood to carve from we would all carve completely different sculptures. The same is true for our tastes in liking or disliking art. Hopefully some of the topics discussed in this chapter will have helped you to decide what makes a good sculpture for you. However, your likes are bound to change as you learn more about the subject, and you can never stop learning and discovering new works of art. Often you will rediscover things about a great work of art, things you might have overlooked the first time, or things that now make sense to you because you have developed.

Sculpture, like all other art forms, is a process of discovery and you can only really understand that process through making the art. *You* will ultimately have to decide what a good sculpture is by looking at your own work and that of others.

PART II
Sculpting in Clay

In this section we will be making a portrait head ready to be cast into bronze. A solid armature is the starting point for creating a rough clay sphere for modelling a head, either from life or from your imagination. Modelling is the process of manipulating the clay by adding and removing it to realize your sculpture.

The completed clay head has then to be transformed into a plaster version, which is achieved by making a waste mould: the clay head is divided into two and then covered with plaster in two halves to make a plaster casing. When the plaster is dry, the two pieces of the mould are removed from the clay head and the original clay sculpture is discarded. The two halves of the mould are then sealed together, and the cavity is filled with plaster: this will be a plaster positive of the original clay head. To reveal this, the plaster mould casing has to be chipped away with a mallet and chisel.

The plaster positive of the head then needs to be made into a wax positive, which will be the vehicle for transforming the original clay sculpture into a bronze one. The wax positive is made by means of a rubber mould. The rubber mould is also in two halves, and each is supported in a plaster case. When they are joined together the cavity is filled with wax to make the wax positive. A rubber mould allows an unlimited number of casts to be made. An edition of bronzes is a limited number of casts from an original.

Making an Armature for a Clay Head

Making an armature for a clay head is a straightforward process because the shape of a head in its simplest form is a ball and therefore requires only a one-prop armature. Thinking back to Chapter 2 and the general principles of armature building, the main consideration when constructing a head is its top-heaviness. You will therefore need to make a very strong central support, because any wobbliness in this support will increase as the weight increases as you add the clay, and this could lead to disaster as your sculpture nears completion.

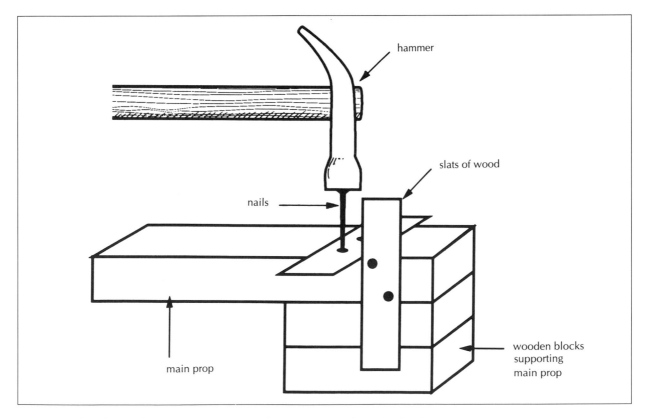

Fixing wooden slats to the top of a central prop for an armature for a clay head.

If you make sure your central prop is firmly attached, however, there should not be a problem. Alternatively you could buy a head modelling armature and stand, but these tend to be expensive.

MODELLING BOARD

For my modelling board I used a piece of 1in (25mm) thick chipboard which I cut from an old shelf to about 12 × 16in (30 × 40cm) in size. Indeed you can use any piece of old or unwanted timber if you want to save money; skips are often good places to find new timber that is surplus to builders' or shopfitters' requirements. Alternatively a piece of 1in (25mm) thick MDF (medium density fibreboard) would be ideal. Whatever you use, it needs to be strong, so an old shelf or offcut from a kitchen worktop is fine.

To enable you to pick up the whole board with ease it is advisable to attach slats of wood underneath. Two pieces of 2 × 1in (5 × 2.5cm) timber nailed or screwed to the board will do the job. If you use screws, drill a pilot hole through your timber first and countersink the end so that the board will sit flush on the work surface.

THE MAIN PROP

There are various ways of making a main prop for a head armature. If you are working within a tight budget you can easily make one using a few old pieces of wood, to save buying expensive aluminium modelling wire. I used an old piece of 2 × 2in (5 × 5cm) timber, about 16in (40cm) long. To this I nailed on some short lengths of 1 × ½in (25 × 13mm) wood all the way round the top of the prop (see diagram on previous page); when doing this you will need to support the head prop so that you can nail the pieces on all four sides. You can then fix your main prop to the modelling board by using four metal brackets approximately 4 × 4in (10 × 10cm); these should be screwed firmly all the way round.

The next step is to bulk out the top of the prop using pieces of packing polystyrene or rolled-up newspaper, held in place by wrapping binding wire around the

Armature for a clay head. This armature was made out of scrap pieces of wood bulked out with polystyrene held in place with binding wire. Metal brackets must be added to this armature to make it sturdy enough to support the weight of the clay.

small slats of wood. It is useful to place a few screws into the prop to tie off the wire. Also wrap short lengths of thinner wire round, to act as keys for the clay to stick to (see photograph). You should bulk out the top sufficiently to fill the inside dimension of your portrait head.

Modelling a Portrait Head

First of all, spend some time considering what type of portrait you would like to do. Will the sculpture be just the head? Will it be the head and neck? Or will you go for the shoulders and bust as well? Sculpting just the face is another alternative. Will the eyes be open or closed; hollowed out or solid? You should also consider what size you want it to be: life-size, over life-size or under life-size. The final cost of the bronze may influence your decisions. In the example here, the artist made the decision to model the head only. The smoothness of the skin contrasts subtly with the light texturing of the hair.

The actual 'look' of the sculpture will depend on your own taste, whether you prefer a very smooth, realistic look or a rough-textured, more expressive appearance. If you intend to tackle a plaster waste mould, it is worth noting that with a smooth finish the plaster casing will fall away more easily.

The amount of detail you include in your sculpture is purely your choice: you have to decide what you, as an artist, are interested in. Ask yourself what you particularly like about the subject before you start. Look

┌─────────── RODIN'S PORTRAITS ───────────┐
Study other sculptors' work and see how they have tackled their portraits. For example Rodin's work encompasses many variations of portraiture: whole busts; head and necks; just heads and faces only. Apparently Rodin often never used an armature when modelling a portrait because he compacted the clay so tightly that it supported itself. He also worked on portraits by laying the head down on its back and supporting it with wedges of clay.
└──┘

critically in order to assess it. You may decide that too much detail will spoil the impact; alternatively you might want to include every wrinkle.

SETTING UP

Take time to set yourself up. Set your armature on a table or modelling stand so that it is at eye level; you need to look up to your sculpture rather than down at it. Placing it on a turntable makes it easier to look at it from all angles with ease. Make sure the space around it is clear so that you can stand back to look while you are modelling.

Modelling tools are useful, and you can buy them in sets with a whole range of head shapes. However, experience has shown that no matter how many modelling tools are to hand, sculptors tend to stick to their favourite two or three; so buying a set is perhaps unnecessary. Also, many sculptors use tools which appear to have no relation to sculpting – an old kitchen spoon, for example. Fingers can also come in handy.

STARTING TO ADD CLAY

Start by pushing chunks of clay onto your armature, working it into the wire; it is important that this first layer of clay 'keys' is held securely by the wire. Continue building up the clay all over the armature, working towards a roughly shaped head.

Look closely at the shape of your sitter's head; walk around the person and try to see the shape of the head under the hair. Build up more clay, but model only very roughly at this early stage. A good way of working is to add just small pieces of clay, pushing them on with your thumb; or you may also find it useful to carve away chunks of clay with one of your modelling tools. You will find your own way forwards with your modelling techniques, as you experiment and gain confidence in using clay.

Adding the first layer of clay. Notice how it has been pushed firmly onto the wire to ensure a strong key.

Adding small lumps of clay and bulking out the form of the head.

At this early stage of your sculpture it is a good idea to take measurements of your sitter's head. Use a 12in (30cm) ruler, or even a length of wood with your thumb as a marker, and measure the length of the head from front to back, the width of the face temple to temple, the size of the forehead, and the overall length of the face from the chin to the top of the forehead. You may wish to use callipers to measure your model's head.

Remember that at this stage you are only trying to establish the rough shape and size of the head, so don't get carried away with the measurements. Keep the sculpture loose, not too precious or refined; it is easy to become engrossed in detail at too early a stage. You can then roughly model the nose and build up the shape of the forehead. Keep adding the clay in smallish

lumps, and don't be frightened of cutting pieces away and rebuilding areas.

WORKING 'IN THE ROUND'

Don't forget to stop and review your work from all angles: it is all too easy to become so absorbed with working on one area that you fail to relate it to the rest of the sculpture. Move round both your sitter and your sculpture frequently; a head doesn't have a front and a back! Working 'in the round' may be new to you and

Establishing the features in their rough form.

Thinking about what is happening at the back of the head while establishing the rough form.

may feel strange at first, but in time you will find your 'sculptor's sense'. Constantly observe your sitter, and manipulate the clay so it is expressive: push it around as much as possible to define an area, using your fists if needs be; cut chunks away if that is what is needed. Enjoy the physical feel of the substance and forget that there may be rules for modelling in clay. *Feel* the clay and enjoy it; make it work for you. Also, at various points in working you may wish to make some drawings of your model so you are more familiar with the form of his or her head.

If you have to leave your sculpture at any point for a long period of time, you should cover it with plastic – supermarket carrier bags or bin liners are ideal – because the clay must always be kept moist.

It takes time really to get into modelling with clay,

but the more you do, the more knowledge you gain for yourself and before long you will be engrossed not just in modelling, but in the subject as well. Look at your sitter's features: the way the nose is formed, how it joins the cheeks, how it flows from the forehead and dips down into the eyes. Look at the hairline if there is one, and see the step from the skin to the hair. Look at the area around the mouth, and notice how from the nose it protrudes out and joins the upper lip. See how the chin is formed. Keep adding and taking away clay to emphasize these planes, ridges and hollows. The more you look with an open and enquiring mind, the more you will understand about the complex and subtle form of the head.

When you have defined the form as much as possible you can start to refine it. If you want your sculpture to have a smoother appearance, then smooth the clay over with your finger tips, dipping them in water as you go along. As you start smoothing the clay you will notice the form of your model emerging. If it's not right, do not hesitate either to build up the form or to cut it away: do not stop this process until you are satisfied, blending the clay in to establish a highly modelled form.

At this stage you may be concentrating more on the features of the face, perhaps looking closely at the shape and form of the eyes. One of your modelling tools could be used to scribe lines or dig away clay. Eyes are so important, and how you actually tackle modelling them depends on you. Once again, look at the work of other sculptors and see how they found a solution. Don't be surprised if you have to try several times to capture the exact expression of your sitter's eyes. A large part of learning to sculpt is learning to look. Do not allow yourself to be swayed by preconception: learn to believe what your eyes tell you.

When modelling a feature, look constantly at how it relates to the rest of the head. For instance, when modelling the mouth observe what happens around it: the space between the mouth and the chin, and how it tapers upwards to the nose. Bit by bit you will have worked around your sculpture and refined it; now cover it and leave it until the next day to have a fresh look at it. You may decide that there are areas you are not happy with, and wish to remodel.

A common problem is knowing how far to take the final stages of developing a sculpture – or, put more simply, knowing when to stop. And of course there is no easy answer; it depends on the sculpture, and on the interests of the artist. The subject itself may also give some guidance in making a decision. If you are

Starting to refine the features.

modelling a child's head a high degree of detail is inappropriate because children have soft shapes and a smooth skin texture. Older people, however, provide an opportunity to indulge in skin and hair textures, and the wrinkles and lines that help build up a person's character. But this must all be thought of within the context of the overall sculpture; you want to go beyond merely recording a person's facial features. With children and young people it can be quite difficult to ascertain the physical character of their faces.

Eventually you will reach a stage when you are happy with your sculpture and will stop working on it. At this point you will need to decide whether you are going to pay a foundry to turn your hard-won creation into bronze, or if you are going to tackle the next stage and make a waste mould. Either way, make sure that your work is well covered up with plastic, as the clay must not dry out.

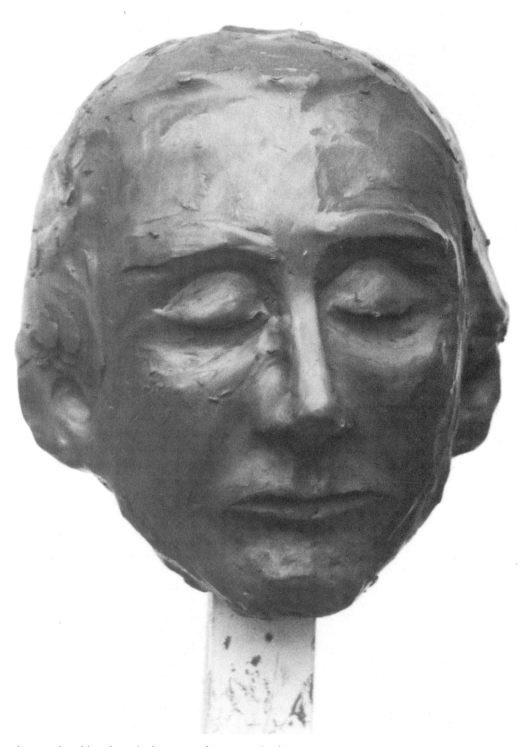

The nearly completed head just before smoothing over the forms.

Making a Plaster Waste Mould from a Clay Head

MATERIALS

- clay head (see Chapter 10)
- clay or brass shim
- scrim
- plaster bowls
- washing-up liquid
- paintbrush
- chisels and mallet
- water sprayer

The purpose of making a plaster waste mould is to enable a plaster positive of your portrait head to be cast: therefore a plaster head instead of a clay head. This process turns an inert material into a durable hard one. Put very simply, the sculpted clay head is covered with plaster, first on one side and then on the other; the plaster halves are then separated and taken away from the clay. Any clay left inside the plaster pieces is removed, and the two halves are fitted back together and sealed with plaster and scrim. This mould is then filled with runny plaster. Once this plaster is dry, the plaster casing (mould) is chipped away to reveal the plaster positive of the head you sculpted in clay.

It is a very labour-intensive process, and if you work swiftly and steadily it will take a couple of days of constant work. It is a messy business, so if you are working inside your house you will need to lay down plenty of plastic sheeting and to wear old clothes. Even if you are a very careful and neat worker you will still find that plaster tends to get everywhere. You will also need to work in an area where you will not be disturbed by children or pets.

THE TWO-PART MOULD

Make sure there is plenty of plastic sheeting both underyour sculpture and covering the area surrounding it. The clay head needs to be divided into two halves so that when the plaster casing has been made the clay can be removed easily. This head is simple to divide up as the shape is essentially a sphere and hence the mould is a two-part one.

First you must decide where to make a division line on your clay head. When the final head is in its plaster form the seam line will show, but this can be smoothed over until it is no longer evident. However, it is not a good idea to have such a line running through, for instance, the middle of an ear, for obvious reasons. First of all, therefore, you have to decide where to place your division line, either placing it to run down in front of the ear or behind it. Due to the shape of this particular head I decided to run the line in front of one ear and behind the other.

To physically divide the head you can use either a clay wall or brass shims. Brass shims are pieces of thin brass sheeting which are cut into wedge shapes; these are stuck into the clay perpendicular to the head so that they overlap each other all the way round the sculpture to make a wall.

Making a Clay Wall

To make a clay wall, roll out some clay (see Chapter 3, the sub-section Rolling Out) using a wooden rolling pin, or you could use an empty wine bottle. The slabs need to be a minimum of ½in (13mm) thick and approximately 1in (25mm) wide by approximately 10in (25cm) long; but if the lengths turn out shorter than 10in (25cm) it doesn't matter. Place the first slab at the bottom of the side of the head and perpendicular to it, pushing gently against the sculpture. Prop it from behind with something heavy and stable, such as a couple of cans of beans or a house brick. Place the next slab on top of this, also pushing it gently against the sculpture. Where the two slabs meet each other, smooth over the ends so that they become one continuous wall.

Continue the wall up and over to the other side of the sculpture, making sure it becomes one continuous wall. Where there are large undulations, on the hair for example, gently crimp the bottom of the wall into the hollows so that there are no gaps between the wall and the sculpture. Remember, however, that the wall should just butt up to the sculpture and should not actually be sealed.

At the bottom of the head or the shoulders continue the wall round, with the wall forming a right angle to the face. You may need to place props underneath this part to support it. Make sure that the wall is stable; it is important that there is no danger of it falling off during the casting process. It is a good idea to flick the clay wall with your fingers lightly to make sure it is sturdy enough; it will need to withstand the same kind of pressure when the first coat of plaster is applied.

Making Mould Keys

The next step is to make what will become mould keys or location pins. Push the handle tip of a wooden spoon or a similar implement into the clay wall to make distinct indentations. Indent at intervals of approximately 3in (7.5cm); but you need not indent the wall at the very bottom of the sculpture. These key marks will later be the keys which will locate the two halves of the plaster casing together in their exact position.

Flicking on the First Coat

You are now ready to apply your first coat of plaster. The first coat needs to be coloured, either by adding to the water a bright poster paint colour, or you could use food colouring or coloured ink. This is so that later on when you chip out the mould you will know when youare reaching the surface of the plaster positive, and you can then proceed with more care.

The finished head showing the clay wall firmly in place, and marking the location pins around the clay wall.

You will need to mix enough plaster to cover the first half of the sculpture with a layer approximately ¼in (6mm) thick. Ideally you should cover all the bare clay on this half in one go, but if you do not manage this, quickly mix another bowl – not forgetting the colouring – and apply it before the first batch of plaster dries out. You have to work very quickly at this stage.

To apply this first coat you literally flick it with your fingers onto your sculpture, working from the bottom up so that no air traps are formed. Take great care to flick it into all undercuts, and with areas such as the nostrils and under the chin make sure you flick up into these recesses, covering all the clay with this first coat. Not all the flicked plaster lands on the head; hence the need for lots of plastic sheeting all around.

Flicking on the first coat of coloured plaster.

The first coat of coloured plaster is nearly complete.

Applying the Second Coat

When the first coat has been applied mix up another bowl of plaster, but without the colouring. Mix it slightly on the thick side – that is, wait for the island to appear in the middle of the bowl and add another three to four handfuls of plaster before mixing it (see Chapter 4, the section Mixing a Bowl of Plaster); then wait approximately five minutes until the plaster goes 'cheesy'. When it is a thick, cheesy consistency hand-build it onto your first coat, working from the bottom up and covering the entire surface of this side of the head. Build up subsequent layers until you have evenly built up the thickness to the height of the clay wall. Make sure that you have covered all of your sculpture with approximately the same thickness of plaster: a protrusion such as the nose will therefore not be too thinly covered.

When the plaster on this first half of your mould has 'gone off', you can remove the clay wall from the sculpture. You will notice how the location pins are now in the plaster positive. If need be, remodel any areas around your sculpture that have been affected by the clay wall and make sure any traces of clay are removed from the new plaster wall.

Mix up a concentrated solution of washing-up liquid with water in a container and brush this liberally onto this new plaster wall with the location pins. This solution will act as a release agent. Repeat this process several times, but make sure that any scum from the liquid is removed because it could affect the plaster on your other first coat.

Now you are ready to apply the first coat to the second half of the sculpture. You will need to build a

Building up the second coat of plaster. This shows the plaster at its cheesy consistency.

The completed first half of the mould, ready to cast the second half. Note the plaster splashes; these should be removed before applying the plaster on the second half to avoid loss of detail.

clay wall across the bottom of the sculpture to contain the plaster, just as you did on the first half. Then apply the plaster in exactly the same way: the first coat coloured, followed by a second coat of a cheesy consistency built up to an even thickness all over.

When you have completed this second half, seek out the join between the two halves. This will be visible as a crack, although you may have to scrape some plaster away to find it.

Separating the Two Halves

When the plaster has gone off you can separate the two halves of your mould. Use some old chisels, or you

could use small wooden wedges. Start by gently tapping one chisel into the crack between the halves. Use another chisel and tap it into the crack further around the mould, and a third chisel (see photograph). As you tap the chisels in, use the water sprayer to squirt plenty of water into the crack all the way around the mould as this will help release the two halves. At this stage you must take care not to tap the chisels too hard or you will risk breaking a section off the mould. The smoother your clay sculpture is, the easier the mould

Separating the plaster waste mould from the clay head. Notice the chisels tapped in all the way round the seam line. Remember to squirt plenty of water into the seam line while undertaking this procedure.

should pull away. And similarly the more undercuts there are, the more resistance will be encountered. Take your time over this procedure; try not to be impatient if it takes a long time.

Remove the mould from the original clay sculpture. This original, together with its armature, is now discarded. Carefully pick away any large pieces of clay left in the plaster casing. If you do have deep undercuts you will have to be careful not to break any of the edges of detail. You can use a modelling tool or piece of wire to pick out any small pieces that are left.

Cleaning and Storing the Plaster Mould

When the worst of the clay is removed, wash your mould thoroughly to remove any traces of clay; an old toothbrush used gently is very helpful here. A hosepipe outside is ideal for washing the mould thoroughly; a shower spray in the bath is second best.

The mould must not be allowed to dry out. It should be kept saturated with water, so if you are not going to carry on with the next stage of casting a plaster head straightaway, keep it wrapped in plastic; this will prevent it from drying out, and it can be stored in this way for as long as necessary. Better still, it should be kept soaking in water in a large container. If you do keep the mould in plastic, then it is important to soak it overnight either in a large container or by pouring water into the two separated halves and leaving it, so the water can be absorbed by the plaster. It is imperative that the mould has been totally saturated with water before you fill it with your casting material.

Casting a Plaster Head from a Waste Mould

Having made your plaster mould, you can now make a plaster positive of the original clay head sculpture.

PROCEDURE

Begin by separating the two halves of the mould and lying them on their backs. Mix up a concentrated solution of washing-up liquid and water, and brush the inside of the moulds liberally; repeat this process three or four times, making sure that any scum from the solution is sponged away. Then gently brush on some vegetable oil, also making sure that any surplus oil is removed. This will act as a release agent and will make it easier for the plaster mould to be removed from the plaster positive of the head when you chip the mould out.

You need to seal the join of the mould. Place the two halves back together, making sure the location pins align correctly. Cut some lengths of jute scrim (you could use hessian or even some old sacking); this needs to be 3in (7.5cm) wide by the circumference of the mould. Mix up half a bowl of plaster and dip the scrim into the wet plaster so it is thoroughly covered and wrap it around the crack of the mould to start the seal. When the plaster goes cheesy you can spread it over the scrim to form a good crust which will join the mould firmly together. The plaster does not take long to go off.

Filling the Mould

The mould is now ready to be filled with plaster. Prop it in an upright position ready for filling: the neck or shoulder cavity opening must be parallel to the work surface, as this is where the plaster is poured in. Lean the mould against a wall or other solid prop, and make sure that it is propped securely on either side as well. When you have filled it you will need to swivel it several times *immediately* to make sure that any air bubbles rise to the surface; obviously you will need to move the side props to do this, and you will need to practise it before you begin so you are used to this procedure and can make sure everything is to hand.

You need to fill the mould with plaster in one go; you cannot top up on plaster which has already gone off. So if you are unsure of how much plaster to mix, then mix a second bucket or bowl of plaster at the same time; it is better to have mixed too much than not enough to fill the mould.

When the plaster is mixed, pour it into the mould as steadily as possible. When this is filled, give it a good swirl round from side to side, trying not to spill the plaster; then give it a tap on the side with a piece of wood. Swirling it around will ensure that the plaster reaches into all the cavities and picks up all the detail of your cast, and tapping it encourages any remaining air bubbles to rise to the surface. If air bubbles were left they would show on the surface of your plaster positive as pock marks – though this is not disastrous, as you can touch them up later.

Leave the mould for a few hours, if not longer, for the plaster to dry out. The next stage is to chip out the mould.

Chipping out the Mould

To 'chip out' is when you use a mallet and chisel to chip the plaster mould away from the plaster positive

Filling the empty plaster waste mould with plaster. The two halves have been sealed together with plaster and scrim; the mould is firmly propped against a firm surface.

inside it. Take care not to get too carried away and damage the head. It is a messy business, so lay plenty of plastic sheeting around the area where you are going to work.

Assuming that you do not have any delicate protrusions on the back of the head – for example a modelled hair ribbon – it is probably best to chip out the back half of your mould first, so leaving vulnerable detail such as the nose until last. By this stage the head will be resting on its back, and any stresses from chipping

out would be taken here which somehow feels less precious than the front of the sculpture. Lay your mould on the work surface and locate the back of the mould (that is, where the back of the head is).

Take a blunt chisel approximately ½in (13mm) wide, and either a wooden or rubber mallet (a rubber mallet is less noisy); then hold the chisel at a right angle to the mould and hit it quite forcefully (see diagram). Repeat this chipping away. As you chip deeper into the mould you will reach the coloured layer of plaster, the first coat you flicked on when making the plaster mould.

Concentrate on chipping down to this layer all over the back half first; you can then tackle the coloured layer bit by bit. Keep the chisel held at a right angle, and tap with the mallet. This layer should fall away quite cleanly. Try not to hit too hard because your chisel may fall onto the surface of your plaster positive and mark it; this is bound to happen in one or two places, though any marks made can be repaired afterwards.

In areas of deep undercuts you will notice that the coloured layer does not fall away so easily, and you will have to tread more carefully. However, once you have exposed one area you will be able to find your way around the detail more easily, and anticipate areas of high detail as you go.

When you have finished the back of the head, turn it over and put an old cushion or pillow underneath to protect your sculpture. Now you can chip out the front in the same way.

When you have completed chipping out the mould you can undertake any repairs and the finishing off. The plaster head will be quite heavy because the plaster will still be wet, so take care lifting it.

FINISHING

You may find that your plaster head is covered in chisel marks; or you may have accidentally knocked off an ear or a section of the hair. This is quite common, and such damage can be repaired. So, if you have knocked a section off – an ear, for instance – then not only should you thoroughly soak the broken part or parts in water, but the area from where it came will also have to be totally saturated. Using a modelling tool, carefully dig out a groove or hollow on both connecting parts.

Mix up a very small quantity of plaster, about a saucerful, and not too thick. Using a modelling tool, fill

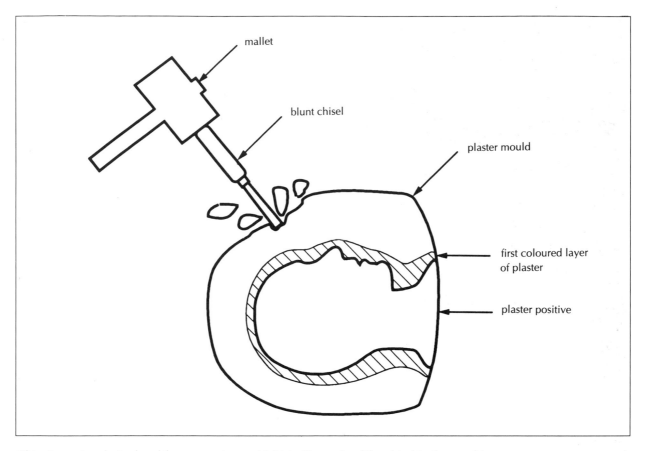

mallet

blunt chisel

plaster mould

first coloured layer
of plaster

plaster positive

Chipping out a plaster head from a waste mould. Note the angle of the chisel to the mould.

this into your grooves or hollowed ends and firmly glue the broken part back on. You may need to spread some plaster along the cracks between parts. As the plaster goes off very quickly you will have to work accurately and quickly.

To fill chisel marks, saturate the area with water and fill them with wet plaster. Smooth over with a modelling tool or a finger.

To finish off: once the plaster is dry, scrape down using either a modelling tool or a fine plaster riffler any flashings remaining or excess plaster around repaired areas.

The Next Stage

The next stage in preparing the sculpted head for casting into bronze is to convert this plaster positive into a wax positive. In order to do this a rubber mould has to be made. If you want the foundry to do this for you, now is the time to ask.

Taking a Rubber Mould from a Plaster Head

Making a rubber mould of a sculpture enables that sculpture to be duplicated as many times as you like. Hence 'editions' of a work of art are taken from one mould; a limited edition puts a market value on a cast from a famous artist.

The purpose here of making a rubber mould is to obtain a wax positive of the original clay head which will then be cast into bronze. But from the rubber mould you could also cast the head in plaster or even concrete. The mould can be used over and over again provided it is looked after.

It is only fair to mention at this stage that because of the complexity of this operation and unless you have had some experience of casting, it is advisable to undertake this process under close technical supervision.

PROCEDURE

The rubber mould is flexible, and a plaster case is always tailor-made to support it. To make a rubber mould is a complicated business, but the procedure is simplified as follows: first, the plaster head is propped face up on a work surface. A dividing line for the two halves of the mould, the same principle as when making the plaster waste mould, is made by using a horizontal clay shelf propped firmly in position from underneath. The face is then covered with a clay blanket; this has what is known as a 'keyway system' around it, a location device to ensure that the rubber mould is always securely held in its case. This clay blanket and keyway will 'become' the rubber positive: it takes up the same space as the rubber while a plaster case is cast over it first.

On top of the clay blanket three solid clay cylinders are placed: these will become the pouring hole and vents. A plaster case is then cast over this clay blanket *et al*, and the head with plaster case is then turned over to allow the clay blanket on the second half to be

made, and the plaster case cast. This case is then moved away, and the clay blanket removed.

The case is replaced, leaving a negative space where the clay was. Rubber is then poured in through the plaster case, filling up the space where the clay was. The mould is turned over, and the process repeated on the second half. You then have two halves of rubber, complete with their mother plaster cases.

TYPES OF RUBBER MOULD

So-called 'flexible moulds' first appeared in the mid-nineteenth century with the discovery of gelatine (elastic animal glue), and made the task of casting much quicker and easier. Before this, sculptures were cast by making plaster piece moulds, a method still used today. The Greeks really perfected the art of piece moulding, sometimes casting a sculpture in hundreds of pieces – hence the name. These pieces were joined together by an outer 'mother' mould made of plaster, and either molten wax, clay slip or gesso was poured in to obtain a cast. Such casts left many seam lines, but these were filed and smoothed away to leave a continuous refined surface.

Obviously these moulds were extremely complicated, but they did allow forms with many undercuts or folds to be cast successfully. Because of their flexibility, rubber moulds today enable similarly complicated forms to be cast in only a few pieces.

There are many types of flexible mould materials available. They are employed in industry, and have many uses other than for casting in sculpture. Most foundries prefer a vinyl hot-pour rubber and various forms of cold-pour rubbers.

● **Latex** is a common form of rubber mould used for most hobby moulds found in art shops. These are simple so-called one-piece glove moulds which easily pull off a simple form with no undercuts.

- **Cold-pour rubbers** are made of silicone rubber or polysulfide rubber, and both are mixed with a chemical catalyst to activate them.
- **Vinyl hot-pour rubbers** have the advantage that they can be used time and time again. There are two grades: hard and soft; for most sculptures it is best to use the soft grade which is more flexible.

When making a rubber mould from an object or sculpture, a release agent should be used to ensure that the rubber mould does not stick to the object; release agents can be purchased from the suppliers of rubber mould materials. If the object or sculpture is made of plaster and is saturated with water a release agent does not need to be used.

This demonstration uses a soft grade, vinyl hot-pour rubber called Gelflex, which is totally non-toxic despite its unsavoury smell. You could use a cold-pour rubber instead: follow the manufacturer's instructions to mix, then follow the same procedure as outlined below. To melt Gelflex, ideally a special melting pot should be used; however, these are quite expensive, and you can melt it just as well in an old non-stick wok with a lid.

MAKING THE RUBBER MOULD

MATERIALS
- clay
- rubber mould: Gelflex
- release agent
- clingfilm
- melting pot

After chipping away the waste mould you have a plaster cast of your portrait head; any damage incurred has been made good (Chapter 12), and a rubber mould can now be made from it. Before you can make the rubber mould, however, you must first make the plaster case which will support it: a rubber mould always sits in a tailor-made plaster case.

Prop your head on the work surface with the face upwards; to make the case it needs to be divided into two halves in the same way as when the plaster waste mould was made, and the dividing line between the two halves should be parallel to the work surface. You can use clay or pieces of wood to prop the head in position, but it needs to be propped very securely as it will be taking a lot of weight.

Start by painting shellac (button polish) on your plaster head so it doesn't get damaged by the eventual hot rubber pour. Then place clingfilm over the face to protect its surface from the clay blanket soon to be made (the second stage of making the plaster case).

Making a Clay Shelf

The first stage in making the plaster case is to build a clay shelf all the way round your division line between the two halves; it should be about 1½in (38mm) wide and ½in (13mm) thick. Because it is horizontal, gravity demands support from underneath: this can be provided by custom-built clay columns or pieces of wood.

Where the neck of the plaster head ends, a hole must be engineered to follow the circumference of the end of the neck or shoulders; the clay shelf should therefore be extended flush over the top of the end of the neck or shoulders. This means the clay shelf at this point will be almost at a right angle to the horizontal clay shelf – thus a semi-circle. When making the wax positive, this hole will enable a core to be inserted before the wax is poured in (explained in more detail in Chapter 14; see diagram overleaf).

Making the Clay Blanket and Keyway

Next you must make a clay blanket – this blanket dictates the thickness of the rubber mould when ultimately the rubber is poured: roll out strips of clay about ⅜in (10mm) thick (see Chapter 3, sub-section Rolling Out) and lay these over the face of the sculpture to cover the entire surface. The strips should be joined together and to the wall by smoothing the clay over to form one continuous surface. The rubber will eventually take the place of this clay blanket: at the moment the blanket occupies the space that will be taken by the rubber mould. When it is poured, the rubber will pick up all the detail from the sculpture so you should not therefore push the clay down into all the nooks and crannies of the sculpture's surface.

When it is finished, the rubber mould will sit in the plaster case for support, and a part of helping it to sit in it snugly is to have a sort of rim known as a 'keyway'. With the blanket complete, the keyway is created next. Roll out evenly sausages of clay about ¼in (6mm) thick; these are placed in the right angle where the clay

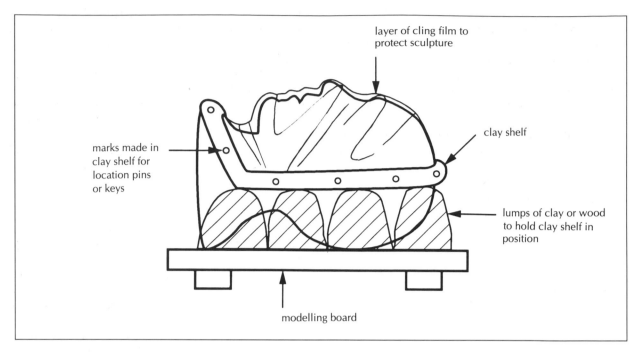

Rubber mould (a) Making a clay shelf and marking location pins.

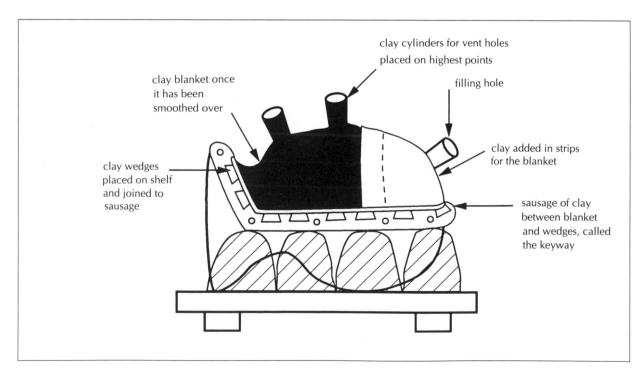

(b) Making a clay blanket with keyway system and clay cylinders for vent and filling holes.

shelf meets the clay blanket, and they sit just to the side of the blanket (see diagram). The sausages need to be joined in such a way that they become one smooth, even, continuous sausage.

Location tabs are made at this stage to ensure that each time the rubber mould is placed in the plaster casing it is put in its correct position. These tabs fit onto the clay sausage keyway. To make them, roll out a clay strip ¼in (6mm) thick, and cut out some wedge shapes about 2 × ¾in (50 × 19mm) long; these are attached to the sausage and the join smoothed. They should be placed approximately every 3in (7.5cm) around the clay shelf, and should not be smoothed to join the surface of the clay shelf; they simply *sit on top* of it.

Mould keys or location pins need to be made approximately every 3in (7.5cm) all around the clay shelf; at a later stage, these will be the keys which will locate the two halves of the plaster casing together in the exact position. Do this by pushing the end of a wooden spoon handle or similar implement into the clay shelf to make distinct indentations. (This is the same as when location pins were made for the plaster case for casting the plaster head in Chapter 11.)

There will need to be holes in the plaster case through which the rubber can be poured into the mould space, and a couple of vents for the air to come out; these are positioned on top of the blanket. The hole where the hot rubber will be poured needs to be at the lowest point on the sculpture: thus on the face it will probably be where the chin is. The holes are created by making three clay cylinders about 2½in (6.5cm) high and 2in (5cm) wide. Position these, and smooth around the bottom to seal them to the clay blanket. The two vent holes need to be positioned on the highest points of the blanket, that is the nose and the forehead.

Making the Plaster Case

All the basic work on the clay mould for the plaster case has been done, so now you can proceed actually to make the plaster case: build a clay wall approximately 1½in (38mm) high around the edge of the original clay shelf; it must be sturdy, because it has to retain the force of the plaster when it is poured over the blanket.

Mix up enough plaster to cover the top half of the mould, and make the mix slightly thicker than normal. Pour this over the blanket, just enough to cover it and fill the clay wall. Let the plaster go cheesy, then cover

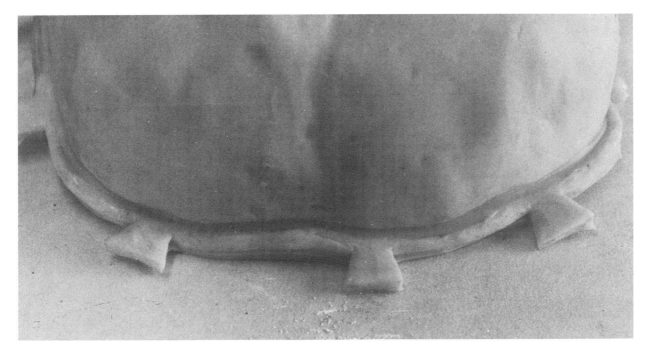

The rubber mould out of the plaster case showing the keyway system.

The rubber mould located inside the plaster case. The location pins can be clearly seen.

the top of the case with the remaining plaster mix, building it up to a thickness of about ¾in (19mm), smoothing it over as you go. Make sure the plaster is thick around the cylinders.

When the plaster has set, you can remove the outer clay wall and trim off the clay cylinders so they are level with the plaster, leaving the clay inside them.

Completing the Second Half

When you turn the mould over, make sure that the plaster head is supported within the case, and take care not to disturb its position; a second pair of hands is very useful here. Prop it all very securely as before, again making sure that the division is parallel with the work surface.

You can now very carefully remove the clay shelf from around the head. The clay wedges should be re-placed should they lift out with the wall; they belong to the first half of the mould. The clay at the bottom edge of the clay blanket may need to be retouched if it has been disturbed by the removal of the clay shelf.

Cover this second half of the head with clingfilm and

make a clay blanket, a keyway and cylinders exactly as for the first half; also complete the semi-circle shelf in the neck area.

Build up a strong vertical clay wall off the side of the now plaster wall (which was the clay shelf) which is on the first half of the case, and prop this up. Wash this plaster surface with a strong concentration of washing-up liquid and water; this will act as a release agent and stop the two halves of the casing sticking together.

Mix up the plaster, pour it and build up this second plaster case in the same way as before.

Putting the Two Halves Together

Make up some metal keys to clamp onto the two halves to hold the mould closed (see Chapter 5, the sub-section Making Keys). Fix these on, leaving the two halves *in situ*.

The end hole needs to be stopped up before the rubber mould is poured; with the mould closed a cap can be made for it now. Stand the mould on its end so that the hole is uppermost and parallel to the floor. Remove the semi-circular clay shelf. You will now be

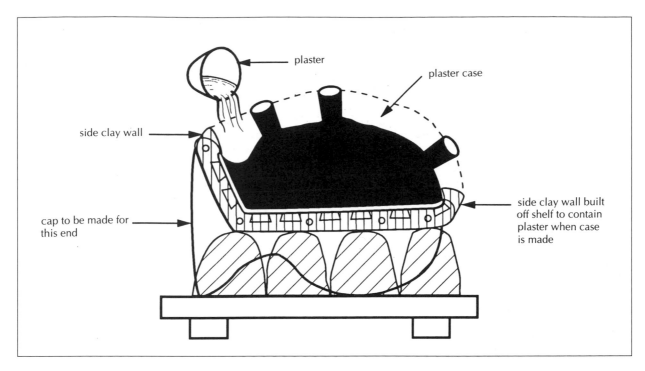

(c) Making the side clay wall for the plaster case and pouring the plaster.

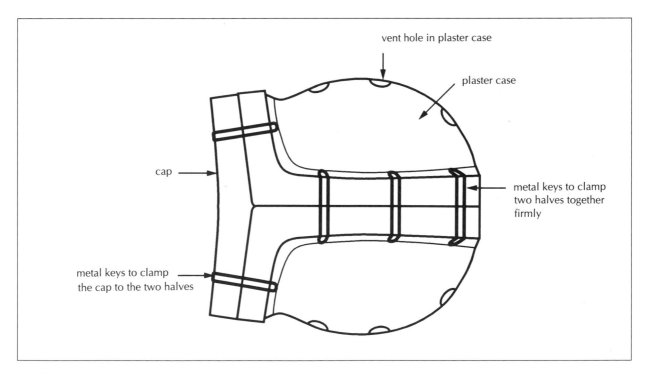

(d) The completed plaster case complete with rubber moulds inside, and the end cap.

looking at the two plaster faces of the end of the mould. Build a vertical clay wall off the side of these plaster faces; it should be about 1½in (38mm) proud of the plaster face, and will ensure the thickness of the cap when plaster is poured in. Now place some clingfilm over the end of the plaster neck. Using the end of a wooden spoon or similar implement, make several holes into the plaster faces of the two mould pieces; these will become location pins for the cap to sit into.

Mix a bowl of plaster and pour it into the hole to fill to the same thickness as the other halves. This cap can be clipped on with the metal keys to the other halves.

POURING THE RUBBER

Lift off one half of the mould and prise the clay out of the plaster case; thoroughly wash out any small lumps of clay left inside.

Remove the clingfilm from the head, and replace the plaster case; seal the halves together by pushing clay along the join. Alternatively, molten wax can be used to seal the cases together. It is essential that the mould is well sealed because when the rubber is poured it could leak out and make a mess, and of course it will not fill properly.

The rubber can now be melted very slowly on a medium heat. It will need to be cut into small lumps and added gradually to the pot. Melt enough to fill the first half of the casing: it needs to be filled in one go. It will take some time to melt the rubber, about one or two hours.

Whilst the rubber is melting you can set up the pouring hole in the case. This is the lowest hole. Take an old baked-bean can and cut out the bottom to make a through-tube; place it over the hole, and seal round the bottom with clay.

The rubber will be ready for pouring when there are no small lumps left in it and it falls off a stirring stick very freely, not slowly. It is extremely hot at this stage, so it is a good idea to wear protective gloves and clothing. Pour it evenly through the filling hole until it reaches the top of the vent holes. It may sink down fractionally after a few minutes, and you should top it up again straightaway.

Leave the mould for at least half an hour before turning it over. The same procedure is now repeated for the second half.

Making a Wax from a Rubber Mould

Your rubber mould and its plaster case are in two halves. Molten wax is applied to the surface of the rubber mould to capture the surface detail. The thickness is built up in layers using a brush to apply the wax. The two halves of the mould are put together, and molten wax is poured in and swirled around to seal the join.

```
┌─────────────────────────────────────┐
│              MATERIALS               │
│  microcrystalline wax                │
│  ½in (13mm) soft paintbrush          │
└─────────────────────────────────────┘
```

PROCEDURE

Make sure your rubber mould is clean; give it a good wash if necessary. Also make sure that the two halves are sitting correctly in their plaster casings.

Melt down half a saucepanful of microcrystalline wax. To pick up the detail of your sculpture, the melted wax is applied to the surface of one half of the mould using a ½in (13mm) paintbrush: do *not* paint it on, just let the wax fall off the brush. On this first layer of wax it is important to avoid trapping air: air pockets mean that detail will be missed, and they will create holes in the surface of your wax positive. You must therefore work as evenly as possible all the way over the rubber, letting the wax fall off the brush.

Build up the layers until the wax is of an even thickness of about ⅛in (3mm). Check to make sure that there are no thin areas in the wax; you must not rush this stage. The wax will need to be thickened slightly round the edges of the mould; excess wax spilling over can be trimmed off with a craft knife.

Repeat the same procedure for the second half of the mould, likewise trimming the edge flush.

Next, put the two halves of the mould together, matching the location pins. Attach the metal key pins and leave the end cap off. Molten wax can now be poured in and swirled around the inside with the aim of following the seam line of the two halves. Keep repeating this until you are satisfied that the two halves are joined together.

Lay the mould down and carefully remove one half of the plaster case, leaving the rubber mould behind; then gently remove the rubber. You can then lift the wax out of the other half of the mould and inspect it: the extent of detail that the wax picks up is always amazing.

The wax head can now be finished by working on the seam line with a warm knife, possibly adding more wax in places. Any other flaws in its surface can be dealt with in the same way. Check the thickness of the wax: because of its translucency, any thin spots will be easy to detect.

Place the wax back into one half of the rubber mould in its case; then resite the other rubber half and in turn its case. The wax is now ready to be taken to the foundry for the core to be poured in.

PART III
Sculpting in Wax

15

Sculpting a Figure in Wax

For the subject of this project I decided to model a ballerina based on a painting by Degas. The ballerina is bending over so the face is not in a prominent position, a point which had a bearing on my choice of pose; and the dress opened the possibilities of shape and texture in the overall structure. You could, however, model any figure in wax by applying the same techniques as described in this chapter, and of those in Chapter 5.

MAKING THE FIGURE

First cut a rectangle of wax from a sheet measuring approximately 2 × 2½in (5 × 6.5cm). The thickness will need to be about ½in (13mm), so you may need to cut two rectangles and join them together, or you could carve a piece from a bigger slab of wax. This rectangle will become the upper part of the body.

Soften the wax in hot water for about ten seconds. Shape one end of the longer side into the shoulders, rounding them. If you are using two sheets, join both edges together by smearing the wax over and fusing it into one piece. The shoulders at this stage are only preliminary and will be bulked out later.

The Head and Neck

Start the neck by cutting a piece of wax from a rod or a sheet approximately ½in (13mm) long, depending on the size of your figure. Use a hot knife to join this to the shoulders by carefully melting the wax around the base of the neck and filling in any areas as necessary. To model the head you can use scraps of wax sheet or rod wax. Soften it first in hot water, then manipulate it into a head shape. You can then use a hot knife to join it onto the neck. Remember the position of the head is looking down, so arrange the plane of the face appropriately.

Roughly form the shape of the hair by adding small

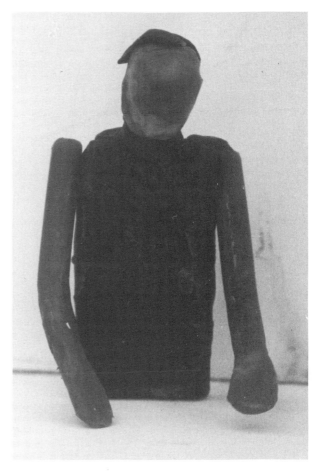

The rough wax torso with arms being added, and small strips of wax bulking out the top of the chest.

strips of wax which have been warmed in water, and arrange them on the head.

The next step is to start the arms. Cut two lengths of ¼in (6mm) rod about 3¼in (8cm) long, warm them in water and model into shape. They do not need to be refined now; you can do this at a later stage.

The Seat and the Right Leg

The top of the seat can be cut from a sheet and measures 4in (10cm) long by 2¼in (57mm) wide. Position the torso on the right-hand end of the seat about ¼in (6mm) from the back and right edge; use a hot knife to join the two pieces together.

Cut a length of ½in (6mm) rod or rolled-up sheet measuring about 3½in (9cm) long for the right leg. Warm it and model it into shape, thickening the area around the knee by adding small warmed lumps and smearing these into shape. Bend about ½in (13mm) above the knee at a right angle; this will be under the dress when it is finished. Use a hot knife to join the leg overhanging the front of the seat.

The Arms and Upper Chest

You can now add the arms, the hands resting in position on the leg; note the position of the arms in the photograph. Build up the areas around the top of the arms, the shoulder blades at the back and the shoulders at the front. Do this by adding small lumps of warmed wax and smearing these into shape. At any point when you are building up with wax, you can use a craft knife to cut away to define forms.

The upper chest can also be built up by adding small lumps of warm wax.

Modelling the Dress

Model and attach the top frill of the dress; this follows the circumference of the figure. Just below this you can judge where the waistline should be, and from this point the front of the dress will slope down towards the right leg. Make this by cutting small rectangles from the sheet wax, and modelling them into an undulating dress shape, arranging them so they overlap one another when placed onto the figure. When adding these you may need temporarily to bend the arms up off the

The wax torso with roughly modelled leg fixed onto the wax seat. The hands have been temporarily positioned on the leg and the hair is being added gradually.

leg in order to attach these pieces initially. These shapes can be built up using strips of wax which are pushed on when warm.

At the moment everything is left roughly modelled; you have been establishing the overall shape of the figure.

Now build the dress around the sides of the figure, overlapping pieces onto the leg as you do so. For the back of the dress you can use one large piece of sheet wax, cut to shape at the top edge and bent to accentuate the flow of the dress at the back. Add this piece by using a hot knife to join it to the torso where it meets at the back waistline, and make sure you obtain a strong weld by placing a strip of wax between the back of the figure and the top of the dress, melting this with a knife between the two surfaces.

Any gaps in the dress at the sides can be filled in with modelled rectangles of wax hot-knifed underneath to make a continuous surface.

The underskirt at the back was made in the same way by modelling a sheet and welding it strongly all round the edges to the underside of the dress. The scallop detail was made by cutting out small triangles of wax and then rounding off the edges using a warm modelling tool.

Fashioning the Left Leg and Finishing the Seat

The left leg can be modelled and added in the same way as the right leg, though you may need to adjust the area between the dress and underskirt to accommodate it. When you are satisfied with its shape and length, you can model and add the ballet shoe.

The back support of the seat can be cut to size by measuring the length of the left leg and the length of the seat. When welding this rectangle of wax to the upper seat it is advisable to melt thin strips of wax into the weld to ensure a strong join between the two surfaces.

Completing the Details

With the overall figure completed and securely attached to her bench, you can now start finishing off areas of detail. The dress can be finished by adding thin strips of wax onto the upper surface and welding these gently into place with a warm modelling tool; these strips can be bent first to follow the contours of the dress. Use a hot modelling tool to scribe in lines

The front of the dress is being established here. The frill round the top of the dress has been positioned; the shoulders have been bulked out. Notice the figure is still in a very rough form at this stage.

around the dress to suggest folds and creases. Finally you can paint on some wax to give a textured finish similar to the quality of fine material.

The hair was smoothed over using a warm modelling tool, and you can fill in any areas that need defining as you go along. The exact shape of the hairline is your choice; I tried to suggest a French plait. Likewise the face can be shaped, first using a hot tool to define the plait and sides of the forehead. The nose can be modelled by adding a small piece of wax and defining it with a warm tool, the eyes and mouth by cutting with a cold tool.

Finally, check your sculpture all over for unfinished detail and any areas that may need redefining, such as the shape of the limbs or any rough marks left by the hot knife. If you wish to end up with a very smooth finish on the figure's body, keep using a warm tool, carefully blending areas over and over again.

Using a warm modelling tool to smooth over the hair.

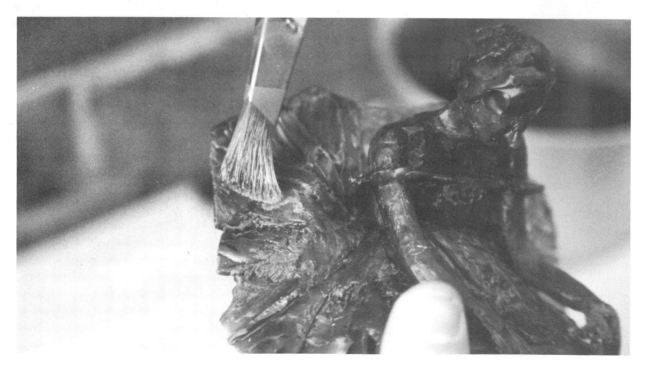

Using a brush to apply wax to the back of the dress to establish the texture of the material. The modelling has not yet been completely refined.

The completed wax figure.

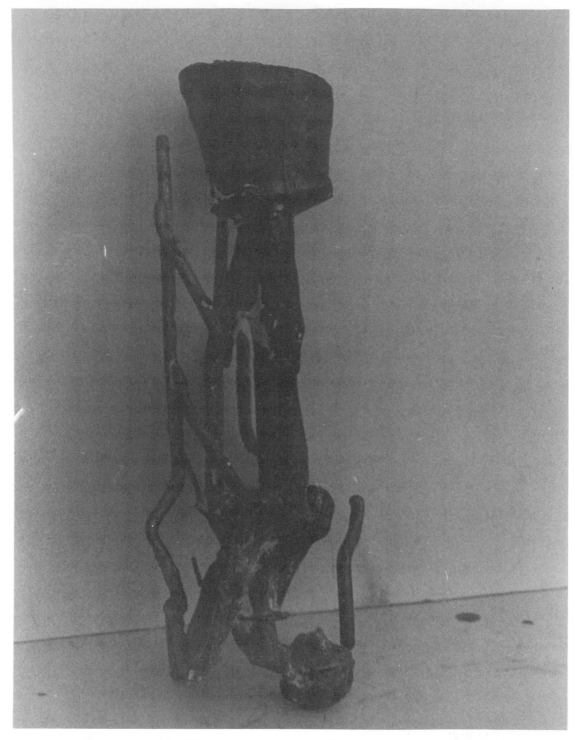

The bronze ballerina showing the filling cup and the runner and riser system. This was cast using the ceramic shell process. The shell has been broken off to reveal the bronze; traces of it can still be seen on the bronze. The pouring cup is at the top; the bronze runners can be seen from the pouring cup feeding into the sculpture in three places: into the foot, into the underside of the bench and into the very bottom of the bench. This is an example of a direct feed, the bronze running into and through the sculpture and out through the risers. A network of risers can be seen connecting at the back of the dress, and the riser from the ballerina's forehead did not completely fill to the top of the mould, hence it being short.

The final bronze ballerina in its completed state after all the runners and risers have been cut off and the sculpture has been chased and patinated; 6½ × 5½ × 2in (16.5 × 14 × 5cm).

PART IV
The Foundry

This part of the book is intended to explain what actually happens to your wax sculpture at the foundry, and how it is cast into bronze. Most artists who take their work along to the foundry have no idea of the process involved in bronze casting; they hand in a sculpture at one end, and as if by magic they collect a nicely finished patinated (coloured) bronze at the other. In between, however, is a process which has many parts to it, and which can go, in some cases, disastrously wrong at any stage. Nevertheless, it is an exciting, magical process, and it is extremely satisfying finally to see your sculpture preserved in the warmness of bronze.

The three main methods of casting bronze are investment, ceramic shell and sand casting. The investment method and the ceramic shell method both use the lost wax process. Other materials can also be cast as long as they are light enough to burn out in the kiln: balsa wood, polystyrene, cardboard and fabric.

- **Investment** This method is described in this chapter.
- **Ceramic shell** A process in which the wax sculpture has runners and risers attached, and is then dipped repeatedly in a ceramic mixture to build up a thin layer, or shell. Because the ceramic mixture is much harder than the investment mixture, the wax need only be covered with about ¼in (6mm) of the mixture; hence the name ceramic shell. Like the investment method, the mould is baked in a kiln upside down and the wax is burnt out and lost, leaving an empty cavity which is then filled with bronze. The moulds are packed into a sand pit for the bronze to be poured in.
- **Sand casting** A process where an object, normally wood or metal, is placed into a sand mixture in a box that is in two halves. The object is buried half-way in one half and the sand in that half is frozen with CO_2 gas. The second part of the box is attached to the first half, and sand is packed into the second half and frozen. The box is then split in two and the object removed, leaving an empty cavity in the sand which can then be filled with molten bronze. The bronze is poured in via a runner and vent system. Sand casting suits forms that are precise, such as geometric shapes with sharp edges. The process is used in industry to cast aluminium alloy engine parts.

16

What Happens at the Foundry

A solid wax sculpture to be cast into bronze should not exceed 1in (2.5cm) in thickness otherwise it will have to be cast hollow utilizing a core (an inner mould); all large sculptures have to be cast hollow, hence a core has to be used. It is possible to cast solid bronzes which are several inches thick, but there is a risk of 'shrinkage'; this shows up as holes left in the surface of the bronze, and is due to the way thick bronze cools down. It can be counteracted by using thicker risers or multiple risers on the wax.

FIXING THE CORE

With the wax head left in position in the rubber mould, the core – a mixture of plaster and grog – is poured in via the cap end hole. Into this core is placed a core vent – either a piece of string covered with wax, or a wax runner – which will allow gases in the core to escape when the bronze is poured. When this has set, the wax head can be removed from the rubber mould and any final touching up of the wax done.

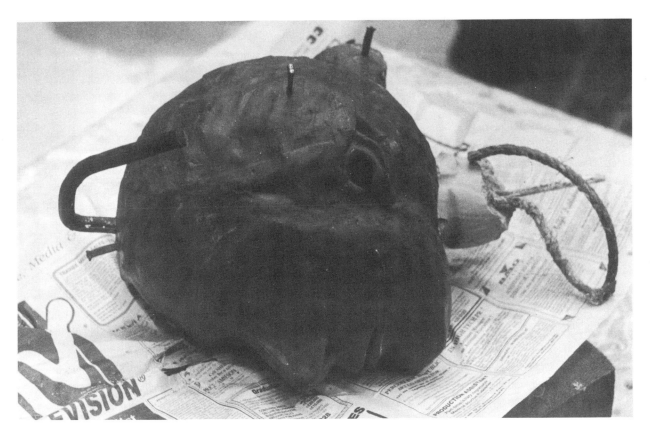

The wax head complete with core inside it. Notice the core vent sticking out of the bottom of the core; and the clearly visible core pins which will hold the core in place. Part of a bottom riser has been placed on top of the head, as the head will be invested upside down.

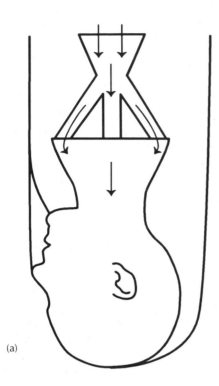

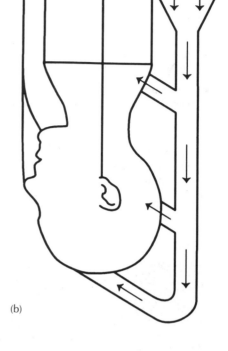

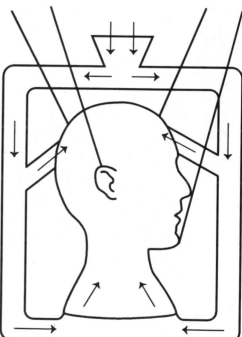

Runner and riser systems for a head: (a) direct feed; (b) step direct; (c) bottom indirect, also known as a slow feed.

Core pins (metal flat nails) are tapped through the wax head into the core to hold it in place when the wax is burned away. The wax is now ready to be 'run' or 'sprued' up.

RUNNING UP A WAX

'Running up' a wax is the term used to describe the attachment of the runner and riser system made of wax rods. When the bronze is poured it is this system which enables the bronze to reach and fill the cavity in the mould which is the sculpture. The runners (thick rods) allow the bronze to flow into the mould, and the risers (thin rods) allow the air and gases to escape.

A runner system has to be designed for each individual sculpture, although a form such as a head can be

The head has been placed upside down on a mould of investment mixture. Stems of risers have been attached to various parts of the head, and the first coat of investment is being painted on the head at this stage.

tackled more or less in the same way each time. Because molten metal is a dense material, turbulence can be caused as it flows through a mould, especially if it is forced to change direction abruptly. The runner and riser system is designed to feed the metal through as smoothly and evenly as possible: if too much turbulence is caused, then the final surface of the bronze may suffer from porosity and have holes in the surface.

Bronze can be fed into a sculpture in three main ways: directly, bottom indirectly, and step indirectly (see diagrams).

● An indirect feed allows the bronze to flow in slowly, filling up from the bottom through to the top, and pushing the gases out through the risers as it does so.
● direct feed fills from the top and flows through a sculpture in much the same way as water going down a plug hole.

The wax head in the photographs has been fed directly at four points on the underside of the chin and neck; note how the runners and risers have been attached to the sculpture using the hot knife technique. The head is set in a base of plaster investment to support it while the system is completed. Once these are attached, a polystyrene cup covered in wax is connected to the runners; this will become the filling cup. The cup, along with all of the rest of the wax, will burn out in the kiln, leaving hollow passages which will allow the bronze to flow into the cavity which will be the head, and allowing the gases out.

The risers are temporarily joined at the top to the side of the cup; these will be freed once the plaster investment is made.

THE INVESTMENT MOULD

With the runner system now complete, the wax head can start to be invested and can receive its first coat of plaster. This consists of a mixture of Herculite plaster (very strong plaster which when brushed on picks up every detail) and fine grog (ceramic material consisting of crushed brick dust and pottery); this is applied very carefully with a brush onto the wax. It has to be laid on as evenly as possible, and built up to a thickness of about ¼in (6mm). To help in the even application of this first coat, air is blown over the surface to dispel any air bubbles in the plaster.

This first layer is crucial in determining the quality of

The wax cup is in position with the thick wax runners attached to the underside of the head. The first coat of investment has been completed.

The risers have been extended up level with the top of the cup; they have temporarily been bridged across to the cup to hold them firmly in place while the rest of the investment is built up.

the surface on the bronze, because any air pockets in the plaster will fill with molten bronze and leave lumps on the surface, known as 'potatoes'. Also, any hairline cracks in this coat will open up when the mould is baked in the kiln and leave flashing marks on the final bronze, known as 'feathers'. In fact it is very common to end up with some form of flashing or infill on the bronze. This can quite often be a shock to unsuspecting artists who first see their bronzes broken out of the moulds – although it is something that you will probably never witness (see Chapter 19, photograph of bronze owl).

When the first coat is completed, a secondary coat is applied which will become the rest of the investment mould; it is a mixture of normal plaster and a material called 'luto'. Luto consists of former investment moulds which have been crushed, so it is essentially recycled bronze moulds. This investment mixture is mixed to a porridge consistency which can be hand-built around the first coat up to the required thickness, normally a hand's width from the risers all round the mould.

Some foundries wrap chicken wire around the mould, with this mixture holding it in place, in order to reinforce the mould. When the investment is completed, the top of the mould is scraped to reveal the tops of the wax risers. This is very important, because when the mould is put in the kiln upside down, the wax will burn out of these riser holes and the cup cavity.

THE KILN BURNOUT

The moulds are stacked into the kiln one on top of

Building up the rest of the plaster investment.

The completed investment mould being smoothed over. The top of the mould will now be scraped off to reveal the riser holes and the cup.

another: the fuller the kiln is, the more economical the firing. Each mould is propped up with pieces of ceramic brick or shelves to allow the wax to run out. Most burnout kilns these days are fuelled by gas, but some foundries may still use hand-built kilns fuelled by propane. In times past, kilns were fuelled by coke and charcoal, the burning fuel being lined around the kiln to enable an even heat to penetrate the moulds.

The moulds are heated initially to a temperature of 400°C. The wax melts out in the first few hours, and this heat is maintained for approximately eighteen hours. The temperature is then increased to 650°C for a period of between six and twelve hours, depending on how big the moulds are.

During this period the moulds can be seen to flame; the flame will burn out of the bottom of the mould, where the cup is, for several hours; this indicates that

there is carbon left in the mould which is burning away, and it is extremely important to watch for. If a mould comes out of the kiln with a black cup or traces of a black cup then it cannot be poured, as any carbon left inside the mould will react with the bronze, causing it to spit when poured. This will lead to porosity and a bad bronze because of holes in the surface.

The final baking temperature of the kiln is 450°C for a finishing-off period of about twelve hours. The kiln is then left to cool for a day. The moulds are still warm when the bronze is poured as this helps the flow of the bronze. Indeed, with a pure copper pour the mould has to be preheated until orange-hot so that this sluggish metal will pour properly.

A kiln full of completed moulds ready for firing.
The moulds have been propped upside down
with a gap between each mould to ensure that
the wax runs out.

SMELTING AND POURING THE BRONZE

The furnace used to smelt the bronze is normally a gas-fuelled one driven by air to blast the heat around the furnace barrel. In ancient times furnaces were built into the ground and were fuelled by charcoal driven by air blown from bellows. Bronze smelts at a temperature of about 1,000°C, but is poured at approximately 1,150°C. It is imperative that every item of equipment is thoroughly warmed before coming into contact with the molten metal; any small amount of moisture would cause the bronze to spit violently, just as a larger amount of water would cause an explosion.

The crucible used to smelt the bronze is preheated until orange-hot, and the first bronze ingots to be used are preheated on the side of the furnace along with the tongs used to pick them up. The ingots can then be gently lowered into the furnace. Each ingot weighs 16½lb (7.5kg), so smelting bronze is a heavy, hot job. A small amount of carbon is added to the crucible to aid the pouring later.

The crucible in the photograph is capable of carrying 100lb (45kg) of molten bronze, which is just about the maximum that two people can safely lift to pour; anything over this weight would be poured with a gantrycrucible. Very large bronzes are poured in several parts, each part being welded together later to complete the sculpture.

When the bronze is ready to pour – the temperature is judged by eye or digital thermometer – the crucible is lifted out of the furnace using a pair of lifting tongs. At this point the heat from the furnace and crucible is almost intolerable, and fireproof clothing must be worn to protect the foundry staff from the heat and from potential spillage.

The crucible is lowered into the pouring shank and tipped slightly forwards to allow a third person to remove the slag (crusty waste deposits) which accumulates on top of the bronze. A degassing capsule is dropped into the pot and plunged, thus causing any impurities to rise to the surface; the pot is again slagged. It is then ready to pour.

The person pouring needs to be accurate and steady; an erratic pour can cause turbulence and therefore poor results. The bronze is poured into the filling cup, and when it appears out of the riser holes the mould is full; it soon dies down to a glowing amber colour as it cools. The sand that contains the moulds is crucial in

The moulds are extremely fragile when they are removed from the kiln, so they are first encased with scrim and plaster to reinforce them. The sand pit containing special moulding sand is dug out, and the moulds are placed in it: it supports them while the bronze is poured; the moulds stick out proud of the level of the sand so that the pourer does not have to bend over too much when pouring.

To ensure that no mould dust or debris will block the runner and riser passages, each mould is blown out with a blowpipe when they are moved. Pieces of weighted-down cardboard are placed over the tops of the moulds to ensure no debris falls down the cup or riser holes while the bronze is being smelted.

containing the force and impact of the metal as it flows through the mould. Bronze cools quite quickly; after about forty-five minutes the moulds can be taken out of the sand pit, and will be ready for opening after just a few hours' cooling time.

The moulds are broken open using a small axe or chipping hammer. The investment falls away revealing a bronze complete with its bronze runners and risers and solid bronze pouring cup. The actual sculpture is wire brushed, and carefully examined before the next stage.

FETTLING AND CHASING

The runner and riser system is systematically removed either by hacksawing close to the surface of the bronze, or it can be cut off using an electric grinder. Any flashing and other casting detritus is removed by using cold chisels, a process called fettling. To speed up this process you can use die-grinders, although in the wrong hands these can do more harm than good.

Chasing is the process of working on the surface of the bronze using small files and matting tools (small metal punches with various textures engraved on the ends). These tools are used to reproduce the original surface where it has been lost to the runners and risers, and to repair other damage incurred in the cleaning-up process. If you examine ancient bronzes it is extremely difficult, if not impossible, to detect any sign of chasing having taken place.

On bronzes which have cores, the core pins have to be removed and the small holes filled; traditionally this task was achieved by drilling the holes and tapping a thread into each one. A small bolt was cut from one of the bronze risers to thread into the hole, the other end being worked over and finally matted to conceal any trace. These days the holes are filled with weld using a

The bronze pouring equipment. The furnace is in the floor with the lid slid across open. The crucible is lifted out using the lifting tongs which can be seen in the middle of the photograph. The crucible is in position in the pouring shank. (This photograph was taken after a pouring had been completed.)

Slagging the molten bronze. The slag is spooned off the top of the crucible and tipped onto a bed of sand. The implement at the very left of the photograph is a plunger used to plunge a degassing capsule into the molten bronze after the slag has been removed.

Pouring the bronze. It is important to ensure a steady, accurate pour. Notice that the other moulds in the sand pit are still glowing as they have just been poured.

The ceramic shell moulds in a sand-casting box. These moulds have just been poured; you can clearly see the thickness of the ceramic shell. The sand box has been used to contain these moulds while they were poured, but it would also be used in two halves for sand castings.

bronze filler rod. The core investment has to be dug out, often having to be soaked out with water.

Finishing a bronze can often be the longest part of the process, and most artists never realize what is involved; they merely collect from the foundry an immaculately finished work of art.

PATINATION AND WAXING

The patination procedure is the actual colouring of the bronze, though before being patinated the bronze is normally shot blasted. This process involves placing the bronze in an airtight cabinet where shot (minute particules of metal or glass) are fired at the bronze by compressed air, clearing any traces of investment left.

The process of patination is a real science, in that there are many formulas and techniques for colouring metals. Most foundries use a brush-on technique which involves warming the bronze with a blow torch and brushing on a chemical solution, slowly building up a depth of colour.

The colour range for this technique represents the most common bronze colours: black, turquoise, orange-brown, dark brown and emerald green. Some of these colours need just one chemical to achieve the patination, others require a mixture of several chemicals. Some of the more complicated patinas involve rubbing on chemicals day by day until a colour is achieved, sometimes using chemicals in conjunction with sawdust, dung and urine.

Once the colour is achieved it needs to settle, and then it must be fixed; normally this is done by applying beeswax to the sculpture. This can be melted in a

The bronze head with its runners and risers attached. The filling cup can be seen with the runners running into the underside of the head as a direct feed and the risers coming from the ears, the nose and the top of the head. Notice the metal core pins attached to the bronze; these will be pulled out and the holes filled.

saucepan and painted on. Bronzes that are sited outside tend to develop a natural patination over the years, a combination of acid rain and birds' droppings giving them that aged Roman green colour.

DECIDING ON A BASE

A base is usually designed to show a sculpture to its best advantage. It can also become part of the overall composition; for example it can accentuate an aspect – if you have a small base for a large figure or vice versa

it can add a whole new dimension to what the sculpture says.

It would be best to decide what type of base you would like before the bronze is patinated, because some fixings may need to be welded on, a process which is best carried out before patination is undertaken.

For a bronze head, a pole made out of a length of runner can be used, one end welded inside the head, the other located in a hole drilled into a base of stone or wood. Alternatively a right-angled bracket can be welded to the bronze and screwed to a wooden base. Sometimes artists attach a fixing plate to the wax, the plate being held by the core when the wax burns away and becoming one with the bronze when it is poured. You may wish to design a fixing plate or pole as part of your wax sculpture.

For wall-hanging bronzes, hooks can be welded onto the back, or holes drilled and threaded to allow small bolts to be inserted which could support a picture wire.

CASTING METHODS TOWARDS A BRONZE

From clay
to plaster mould
to plaster positive
to rubber mould
to wax
to investment mould, ceramic shell
to bronze

From clay
to plaster mould
to wax
to investment mould, ceramic shell
to bronze

From clay
to wax
to investment mould, ceramic shell
to bronze

From plaster, stone, fired clay, metal
to rubber mould
to wax
to investment mould, ceramic shell
to bronze

From wax, found objects, mixed media
to investment mould, ceramic shell
to bronze

From wood, metal, stone
to sand mould
to bronze

The final bronze head. The core pins have been removed, the runners and risers have been cut off, the surface has been chased, and the plaster core inside the head has been removed. The head is patinated and mounted; 9 × 11 × 7in (23 × 28 × 18cm).

PART V
Projects

17

Making a Clay Relief

This project involves the use of three techniques: modelling clay, taking a plaster cast, and casting wax from the plaster. It is a simple process and has the advantage that you can re-use the plaster mould as many times as you like. You could tackle any size of relief for this project, but obviously the bigger it is, the more expensive it will be to have cast into bronze. The size of the relief in this example is 5 × 4½in (13 × 11.5cm).

The most important thing to remember when modelling a relief is to avoid deep undercuts in the clay, because when the wax is cast it needs to release itself easily from the plaster mould. Bevelling the edges of the relief slightly will allow easy release from the plaster.

MODELLING A CLAY SHEEP

The first step is to roll out a slab of clay as described in Chapter 3, on plastic sheeting on a board; the thickness should be about ¼in (6mm). Decide upon your subject matter for the relief. The artist here decided to depict a lone sheep. Start adding small lumps of clay to the clay slab, pressing these over one another, roughly mapping

Establishing the rough form of the sheep.

Using a modelling tool to define the detail of the sheep's coat.

The completed clay relief inside the clay wall ready for the plaster to be poured.

out the form. Remember that modelling is a process of adding and taking away clay; and even though a relief is not a three-dimensional form the same principle applies, so use your modelling tools to cut away unwanted sections of clay.

With the overall design established you can now start to define its constituent forms and to work on details such as texture. Here, the sheep's coat was modelled rough and deep, whereas in contrast the head and legs were made smooth. Use your tools to scribe in details such as the mouth and grass; wet your fingers if necessary to smooth areas over. Remember to avoid deep undercuts so that when a cast is made it will release with ease.

In the example shown the sheep's coat is quite highly defined in places, although the forms are rounded to allow release. Think of the bevelled edge around your relief, and apply that principle to your forms.

MAKING A PLASTER CAST MOULD

Build a strong wall around your relief, with at least a 1in (25mm) gap between the relief and the wall. The

Pouring the plaster into the mould. Notice that the clay wall is surrounded by pieces of wood to hold it in place whilst the plaster is being poured.

wall can be a clay wall reinforced with wood, it can be made from wood reinforced with clay, or it could be a very sturdy wall made only from clay. But whatever its material, it must be strong enough to stop the force of the plaster spilling out when it is poured.

The height of the wall will depend on how high your relief is. Hold a ruler next to your relief, and from the highest point on your form, measure upwards by approximately another 2in (5cm). This will give you the measurement for the height of the clay wall.

Mix up enough plaster to fill the mould and pour it evenly over your relief, leaving about a ½in (13mm) gap from the top of the wall.

Allow the plaster to set, then remove the wall and turn the mould over: the clay relief should pull out easily, leaving only small amounts of clay behind in the details. Wash the mould thoroughly with water, removing any clay left.

RELIEF SCULPTURE

Relief is a type of sculpture that dates back to ancient times. It is rooted in architecture when sculpture began to be an integral part of a building, and works on the premise that it will be viewed vertically from one main viewpoint, that being from the front. In reliefs sculptural forms rise from a flat surface. There are many examples of relief in architecture to be seen, as a walk in any city centre will reveal, in buildings new and old, in museums and in books. Drawing from reliefs will help you to see possibilities for making your own. The relief could be bas (low), when there is hardly any height to the forms depicted; or it could be high, where forms stick out generously. It would be worth looking at both Egyptian and Greek reliefs to note the differences.

The plaster mould of the sheep after the clay has been removed. The sheep is in its negative state.

The wax positive of the sheep. Any final trimmings can be undertaken at this stage.

TAKING A WAX CAST

While the plaster is still very wet you will be able to pour in molten wax. Melt the wax as described in Chapter 5. Before pouring it into the mould make sure there is no surface water on the plaster. Pour in the wax when it is not too hot, then leave it to cool.

As the wax cools, the wax relief will release itself around the edges. After about half an hour it can be carefully prised clean of the mould. Any final adjustments can now be made to the wax. Alternatively you could change your relief by remodelling parts.

In the second example shown here, the artist has taken the wax cast a stage further with the addition of a simple organic found object. The bird's tail is made up of conifer leaves which were fixed on by brushing wax over them. This use of the found object gives the final bronze a very rich textured finish.

Feel free to experiment in this way, adding objects that can be burnt out in the kiln with the wax.

A clay relief of a bird, made using the same procedure as that used for the sheep.

Pouring wax into the plaster mould of the bird. Remember that the mould should be wet but without any surface water, and the wax should not be too hot when it is poured.

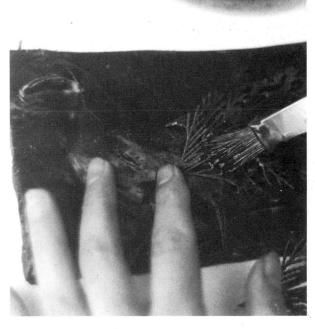

Adding foliage to make the bird's tail, fixing it in place by brushing on molten wax.

The bronze sheep relief with its runners and risers attached. The runners coming from the filling cup are feeding into the back of the relief in three places; this is an example of a step indirect feed. The risers can be seen at the front of the relief, coming from the bottom and the top sides of the relief.

The final bronze sheep relief by Barbara Cheney; 5½ × 4½in (14 × 11.5cm).

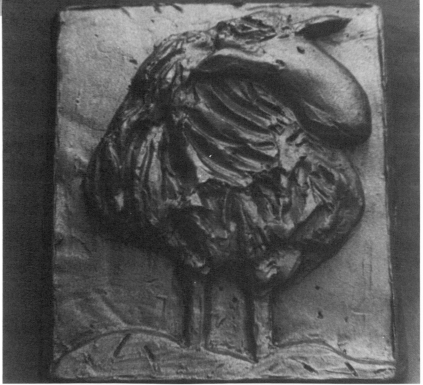

The bronze bird relief showing its filling cup and runners; an example of step indirect feed. The risers broke off during the chipping off of the ceramic shell. This often happens to risers in both ceramic shell and investment mould processes.

Front view of the final bronze bird relief by Barbara Cheney; 7 × 4in (18 × 10cm). All that remains to be done is for the cup and runners to be cut off.

Making a Wax Relief using a Clay Press Mould

This project is very direct and fun to make. Unlike the clay relief project which made use of a clay positive being cast into plaster and then into wax, here we will be using a clay negative and pouring wax *directly* into it. The fun part of this method is that you never quite know what you may end up with: you can achieve some unusual results by experimenting with various objects and materials to obtain some rich textures and effects in the clay; you could even make a complete wax figure by using this technique – dig the clay out with a modelling tool or similar implement and use a smooth or rough object to imprint the clay. Working in the negative in this way is quite a different experience from modelling or constructing in the positive.

If you use a very large bed of clay, say two whole bags of clay shunted together, you could make a large abstract wax sculpture to be cast into bronze. By digging into the clay, texturing it, pushing in blocks or sheets of wood, steel objects, pine cones, fabric material and parts of your body, a creative sculpture can be made. You take a lot of risks by working with this method, but you *can* also control the forms to a large extent.

Molten wax can then be poured into the sculpted cavities and left to cool. When it has cooled, gently push the clay away from the wax and dig away from the wax forms. Great care needs to be taken around areas which are very thin, though any damage incurred can be repaired afterwards.

When the wax sculpture has been revealed, you can then consider it. It could be added to or cut down, or you may even wish to cut the sculpture up and re-assemble it in a different order.

MAKING THE RELIEF

Decide what size you would like to work in, and prepare a bed of clay. I used a freshly opened 27½lb (12.5kg) bag of clay, and the final size of my relief was 10 × 9in (25 × 23cm), so I did not constrict myself. Obviously the bigger you make the relief the more expensive it will be to have cast into bronze, but you should not let that inhibit you.

Choose the subject matter for your relief; you could base this on a famous painting, a drawing you have done or a view of a landscape. Perhaps you may wish to attempt a self-portrait in the negative. If you plan a composition, remember that the cast will be reproduced the other way around from the clay impressions: what reads left to right in the clay impression will read right to left in the plaster cast.

I approached this relief as an abstract composition, although the final form is quite figurative, and used various modelling tools to carve away the clay and to smooth areas over. Fingers come in useful to smooth, push and pinch areas. The decorative lines near the top left were made using the edge of a piece of corrugated cardboard.

Ideally the final thickness of the relief should not exceed ¾in (19mm), although some areas of high relief may be substantially thicker than others; this should not, however, pose a problem for the casting process.

A bed of clay showing the indented and modelled surface before the wax was poured.

The wax positive, demonstrating a variety of marks and textures and unusual forms. Note the lines at the top left hand corner made by the edge of the corrugated cardboard. Below this is a texture which was made by pressing a piece of scrim into the clay 10 × 9in (25 × 23cm).

Making a Bird out of Wax and Found Objects

Here I collected together some found objects and two pieces of fruit to make a bird (see photograph); although you can, of course, use anything you like provided the objects are not too thin. If you decide to use a very thin form such as paper, leaves or cloth, remember to paint on some wax in order to thicken it. On this bird fern leaves have been used which are extremely thin; by painting on wax and building up the thickness, however, any exposed leaf edges should fill with bronze, although one does take a risk by leaving any parts very thin – but it is certainly a risk worth taking for the finished object.

A range of found objects has been incorporated in the making of this bird, including an apple, a lemon, dried poppy heads, fern leaves, fir tree leaves and a dried artichoke stem. You can use anything that you feel is appropriate, such as, for example, real birds' feathers.

The fun part of making a sculpture like this is in finding and inventing objects to use. It is a wonderful

The materials used to make the owl. Notice the sheets of wax in the background.

opportunity to use your creativity to collect objects together and play around with them, finding new ways of making things and giving objects new meanings. It is exciting and fun, so do enjoy yourself and experiment as much as possible.

MODELLING AN OWL

Fashioning the Head

It only occurred to me at the last minute to use the apple for the owl's head. The first step in making the owl is to cut the apple in half: any size apple will do; this one measured about 2½in (6.5cm) in diameter.

Then take the two poppy heads, and using a sharp craft knife, cut around the top of each poppy head just below the frill. Cut about ⅛in (3mm) below where the frill ends. Place the frills on the apple for the bird's eyes; when you are satisfied with their position you can attach them. Cut some small, thin strips of wax from your sheet or other scraps of wax, and using a hot knife or modelling tool, melt the wax around the frill thereby

The owl's head: half an apple, two dried poppy frills and a wax beak.

welding it to the apple. Due to the juice in the apple, there will be some spitting as you melt the wax onto it; you may also find that the wax does not stick very well to the apple. If this happens, keep trying: the moisture from the apple will soon start drying out, and you will find that shortly the wax will stick well.

When both eyes are stuck onto the apple, cut a beak shape from the sheet of wax: the precise size depends upon the personality you wish to give your bird. To stick it to the apple, follow the same procedure as for the eyes, cutting very thin strips of wax and using the hot knife to weld them to the beak and apple. If you slip with your hot knife and burn too much wax away from the beak, cut a thin strip or square of wax and place it over the damaged area; then carefully melt it in, using the hot knife to smooth over the wax, to get your original form. After a little time you will soon get used to using the knife and controlling the wax.

Making the Body

Once the head is complete (see photograph), put it to one side and start making the body: cut four rectangles of wax from your sheet measuring approximately 4 × 3in (10 × 7.5cm). Place the four pieces into a bowl of hot water and leave for a few minutes to soften; when the wax is malleable enough – not too gooey, though – stack the four pieces together and squeeze them around the edges, modelling the wax as you go so that the rectangle becomes one piece of wax. Use your thumb to achieve this by just smearing the wax over from one edge to another, thus sealing the edges.

Once you have done this, model the wax at one end to achieve the shape as shown in the diagram; you may have to soften the wax in the water for a few more minutes. When you have arrived at the shape of the body, you can cut it to make the two wings: take a craft knife and cut through the layered wax as indicated in the diagram. You should make the cut to almost half-way up the body, though make sure you cut right through the wax, exposing the four layers as you do so.

Now, resoften the body in the water and carefully bend back each outer wing; bend them back as much as you like, and as you do so you will notice that the layers of wax appearing on the inner side of the wing take on the appearance of feathers. The wax should bend easily if it is softened sufficiently. However, if it does split as you bend it, put it back in the water to resoften it and then bend it again, smearing the soft surface to seal the split.

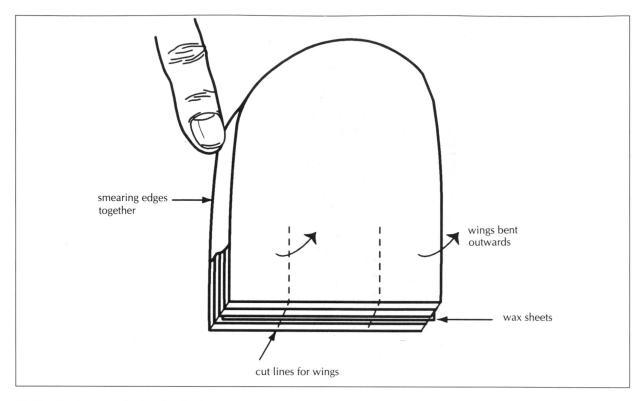

Making the body and wings for the bronze owl.

Putting on the Legs

You are now ready to put on the legs. Dried artichoke stems were used here because of their thickness and their fluted quality, but you may find these difficult to obtain. Essentially you can use any dried flower stems as long as they are thick enough to support the wax body. You could use twigs or pieces of wood, anything with some interesting detail. It's fun trying to find something novel and interesting.

The legs should be about 3in (7.5cm) long and about ½in (13mm) thick. Take the body and cut a shallow 'V' recess into the bottom of it for the leg to sit into. Use a hot knife to slightly melt the wax around the leg to join it to the body. At the top of the leg melt in a small blob of wax to fill in the gap between the leg and the underside of the wing. You will have to use some small sausages of wax and melt these around the thighs to join the legs firmly to the body. When the legs are securely fixed to the body you will be ready to add the foliage.

Adding the Foliage

First you will need to melt a 3in (7.5cm) square block of wax in a saucepan. Remember to heat it slowly on a low temperature, and not to leave the saucepan unattended. When the wax is in a liquid state you can prepare your pieces of foliage. The wax should not be too hot: to test it, dip in your brush: if the bristles burn it is too hot. Fern and fir tree leaves were used, but you may like to use something else. You could use real birds' feathers, for example: the final bronze will pick up all the detail from a bird's feather. Lay the feather on the surface of the wax, and fix it to the body by pushing it lightly into the surface and painting very lightly a lick of liquid wax at its top and bottom ends. You could build up several layers of feathers by using this method. This is precisely the method used to secure the foliage to the body: holding each piece in position and brushing a thin amount of wax over it to secure it. Repeat this all around the body until quite a thick layer is achieved, bulking out the breast of the bird until it reaches the final rounded form.

Fixing foliage to the wax body by brushing on molten wax.

Ideally you should only allow the edges of the leaves to stick out of the body in a few places; keep painting on the wax to flatten them down as much as possible because there is always a risk that the very thin areas of leaves will not come out in bronze. However, it is worth leaving thin areas exposed in a few places on the body in the hope that they will come out in the final bronze. In the photograph you can see that the detail of the fir leaves has come out in its entirety.

Fixing on the Head

When you have completed the body, the next step is to fix on the head. To do this, simply take a 2½in (6.5cm) long piece of dried artichoke stem and push it into the apple, and then into the top of the body. Obviously you have to think about the positioning of the head in relation to the body. This bird's head is slightly tilted to one side to give it a cheeky pose. You could use a piece of dowelling or half a pencil to fix the head to the body.

To finish it off, take a few more small sausages of wax and use a hot knife to join them around the bottom of the head to make a continuous form between head and body.

For the base half a lemon was used, but you could use half a large apple or other fruit, or anything of your choice which seems appropriate. Pierce two holes slightly smaller than the width of the legs, and push the legs about 1in (2.5cm) into the lemon; seal them with a small sausage of wax around the hole using a hot knife. The bird should be able to stand without wobbling at all. If it does wobble you may have to push the legs into the fruit a little further.

Once you have completed the owl, put it carefully into a plastic bag and then into the freezer compartment of your fridge; this will help preserve the fruit until you are ready to take the bird to the foundry to be cast. Also, cooling the wax ensures that all the components of your sculpture remain firmly fused together.

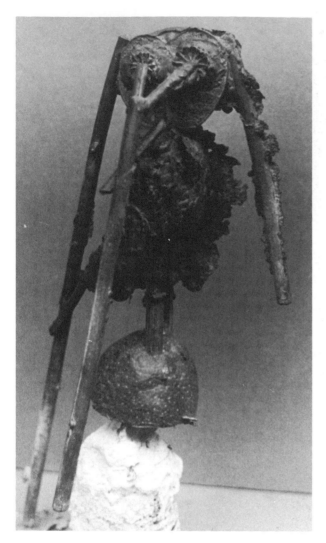

The bronze owl after it had been poured, complete with bronze runners and risers. The pouring cup is at the bottom still covered with plaster; this was a direct feed, the runners going into the lemon. Note the thin line of flashing around the owl's body, and the riser system coming out of the head.

A detail of the foliage on the bronze owl.

The final bronze owl; 9 × 3 × 2½in (23 × 7.5 × 6.5cm).

Glossary

Alginate-hydrogum   A form of rubber that comes in powder form, used mainly for life casting and dental casts.

Armature   A structure made of wood, metal or wire to support clay or plaster.

Back iron   A metal support attached to a modelling board used to support wire or wood armatures.

Banker   A very strong, tall table used for carving.

Beeswax   Natural, refined wax used as a modelling wax or mixed with turpentine to make beeswax polish.

Brass   An alloy of copper and zinc.

Bronze   An alloy of copper and tin with minute proportions of lead and zinc.

Burnout   The flaming of a mould in the kiln; the wax deposits are burnt away.

Callipers   A measuring instrument.

Caps   Sections of a mould which are designed for emptying the mould of clay and filling with casting material.

Cast   A sculpture made from a mould.

Ceramic shell   A method of bronze casting using ceramic material to coat the wax with a thin layer or shell.

Chasing   The process of tooling the surface of bronze, using files and matting tools, to obtain the original surface lost in the casting process.

Chipping out   The process of removing a plaster waste mould using a blunt chisel.

Cire perdu   Meaning 'lost wax', as in the lost wax process of bronze casting.

Clay   Sticky natural substance, a form of soil which in its pure form is composed of the dust from igneous rocks combined with water.

Concrete   A mixture of cement, sand and aggregate.

Core   The inner mould of a hollow metal cast composed of plaster and grog.

Core pin   Also known as chaplets, they can consist of thin wire, nails or thick bars of metal in a massive bronze, used to hold the core in place in the mould.

Crucible   Ceramic container used to melt metals.

Expanded metal   Metal mesh used as a reinforcing material for plaster or cement.

Feathering   See Flashing.

Fettling   The process of removing any casting detritus left on a bronze sculpture.

Firing   The process of baking in a kiln.

Flashing   Also known as 'feathering'. Thin sheets of bronze often left around a cast due to bronze running into cracks in the mould.

Found objects   Ready-made or natural objects that can be incorporated in a sculpture.

Gesso   Plaster made from gypsum and whiting.

Going off   The point at which plaster starts to set hard.

Grog   Crushed brick dust or pottery added to clay or plaster.

Gypsum   A mineral which is the main constituent of plaster.

Hot knife   Welding wax forms together using a heated knife or tool.

Invest   Covering a wax sculpture with a plaster mould in the bronze casting process.

Keys   Location pins to secure plaster moulds together.

Keyway   A system of location devices to secure a rubber mould in its plaster case.

Kiln   An oven for firing clay or moulds to a high temperature.

Latex   Rubbery material used for mould making.

Life size   The size equal to the actual living size.

Location pins   See Keys.

Lost wax   See *cire perdu*.

Luto   A material composed of recycled plaster investment moulds.

Maquette   A small sculpture or study made in preparation for a larger sculpture.

MDF   Medium density fibreboard, a form of hardboard.

Mod-roc   The trade name for plaster bandage offcuts.

Mould   A negative shape from which a cast can be made.

Papier mâché   Material composed of pulp paper and offcuts used with glue.

Patina   A colour which can be applied to bronzes and other metals using chemicals.

Piece mould   A mould assembled out of separate pieces, often of many parts; generally made of plaster.

Plaster   Material in powder form which when mixed with water becomes hard. Made from gypsum and used for mould making and can be used directly.

Plunger   A steel tool which is used for degassing molten metal.

Positive   The final cast of a sculpture.

Release agent   A substance used to facilitate easy release of a material from a mould. Also known as a 'parting agent'.

Relief   A picture with sculptural qualities pertaining to the depth dimension.

Rifflers   Tools with fine filed surfaces used on stone, plaster and metal.

Riser   A channel that allows gases and air to escape from a mould when molten metal is poured. Also known as 'sprue', 'air' and 'vent'.

Runner   A channel through which molten metal is poured into a mould from a filling cup.

Sand casting   The casting of metal in sand whereby the compressed sand acts as a mould. Only suits geometric shapes.

Scrim   A fabric usually made from jute used to reinforce plaster.

Shellac   A varnish made from the resinous secretion of insects found on trees in south-east Asia.

Shim   Thin sheets of brass used to divide a waste mould.

Slag   The crusty waste deposits left on top of molten metal.

Slip   Liquid clay used for sticking clay pieces together.

Sprue   See Riser.

Surform   Tool used to smooth and refine plaster.

Undercut   The part of a form that goes under and which would not release itself easily from a mould.

Vent   See Riser.

Waste mould   A mould from which only one cast can be taken due to the mould being broken away to reveal the positive form.

Whiting   A mineral mixed with animal glue.

List of Suppliers

Some suppliers for tools and materials may be found in the following art publications. If you have difficulty finding a suitable supplier, contact your local arts society, or university/college which runs an arts course; they should be able to suggest the best suppliers in your area.

UK
Artists' Newsletter
PO Box 23
Sunderland SR4 6DG
Tel: 0191 564 1600

Crafts
12 Waterloo Place
London SW1
Tel: 0181 839 6306

USA
Art Forum
65 Bleeker Street
New York 1003

Art News
PO Box 969
Farningdale
New York 11737

Arts Magazine
23 East 26th Street
New York 10010

Canada
Vie des Arts
Rue Francois Xavier
Montreal
Quebec

Australia
Art and Australia
653 Pacific Highway
Killapa 2071
New South Wales

New Zealand
Art Magazine Press
PO Box 7008
Auckland
New Zealand

Europe
Parkett
Quellenstrasse 27,
8005 Zurich
Switzerland

Nike
Postfach140540
D-8000 Munich 5
Germany

Index